Issued under the authority of
(Fire and Emergency Planning)

Fire Service Manual

Volume 1
Fire Service Technology, Equipment and Media

Firefighting Foam – Technical

HM Fire Service Inspectorate Publications Section
London: The Stationery Office

© Crown Copyright 2000
Published with the permission of the Home Office
on behalf of the Controller of Her Majesty's Stationery Office

Applications for reproduction should be made in
writing to The Copyright Unit, Her Majesty's Stationery Office,
St. Clements House, 2–16 Colegate, Norwich, NR3 1BQ

ISBN 0 11 341188 X

Cover photograph: The Fire Experimentation Unit

Half-title page photograph: West Midlands Fire Brigade

Printed in the United Kingdom for The Stationery Office
TJ763 4/00 C50 5673

Firefighting Foam – Technical

Preface

This manual, Volume 1, Fire Service Technology, Equipment and Media – Firefighting Foam, deals with technical aspects of foam concentrates, standards and equipment.

This book complements the existing manual in Volume 2 – Fire Service Operations – Firefighting Foam.

These books replace:

> The Manual of Firemanship Book 3, Part 3
>
> Dear Chief Fire Officer Letter 2/97 – Foam Application Rates.

The Home Office is endebted to all those who have helped in the preparation of this work, in particular:

> Mr Bryan Johnson BSc.;
> Home Office Fire Experimental Unit;
> Mid and West Wales Fire Brigade;
> Angus Fire Armour Ltd;
> Williams Fire and Hazard Control Inc.;
> Civil Aviation Authority;
> British Fire Protection Association Ltd;
> Cheshire Fire Brigade;
> London Fire Brigade;
> Fire Service College;
> Dr Tony Cash;
> Northern Ireland Fire Brigade.

Home Office, April 2000

Firefighting Foam – Technical

Contents

Preface		iii
Chapter 1 Introduction		**1**
1.1	General	1
1.2	Historical Development of Firefighting Foams	2
1.3	How Foams Extinguish Fires	3
1.4	Production of Finished Foam	3
1.4.1	General	3
1.4.2	Percentage Concentration	3
1.4.3	Aspiration	4
1.5	Foam Expansion Ratios	5
1.5.1	General	5
1.5.2	Equipment Used For Generating Different Expansion Ratio Foams	5
1.5.3	Foam Concentrates	5
1.5.4	Typical Uses and Properties of Low, Medium and High Expansion Finished Foams	6
Chapter 2 Foam Concentrates		**7**
2.1	Types of Foam Concentrate	7
2.1.1	General	7
2.1.2	Protein Based Foam Concentrates	8
	(a) Protein (P)	8
	(b) Fluoroprotein (FP)	9
	(c) Film-forming Fluoroprotein (FFFP)	9
2.1.3	Synthetic Based Foam Concentrates	9
	(a) Synthetic Detergent (SYNDET)	9
	(b) Aqueous Film-forming Foam (AFFF)	9
2.1.4	Alcohol Resistant Foam Concentrates (AFFF-AR and FFFP-AR)	10
2.1.5	Hazmat Foam Concentrates	11
2.1.6	Wetting Agents	11
2.1.7	Class A Foam Concentrates	12
2.1.8	Fuel Emulsifiers	12
2.2	Handling and Storage of Foam Concentrates	12
2.2.1	Compatibility	12
2.2.2	Viscosity	13
2.2.3	Corrosion	14
2.2.4	Storage and Use Temperature Conditions	14
2.2.5	Order of Use	15
2.2.6	Storage Containers and Bulk Storage	15

Chapter 3 Foam Concentrate Standards and Periodic Testing   17

3.1	General	17
3.2	Physical Property Tests of Foam Concentrates	19
3.2.1	General	19
3.2.2	Specific Gravity (Relative Density)	19
3.2.3	pH (acidity/alkalinity)	19
3.2.4	Sediment (Sludge)	19
3.2.5	Spreading Coefficient	19
3.2.6	Effects of Freeze/Thaw	20
3.2.7	Accelerated Ageing	20
3.2.8	Viscosity	20
3.3	Foam Concentrate Standard Fire Tests	20
3.3.1	General	20
3.3.2	Is the Fuel Commonly Encountered Operationally?	20
3.3.3	Is the Fuel Reproducible?	21
3.3.4	How Long is the Preburn?	21
3.3.5	How Deep is the Fuel?	21
3.3.6	What is the Application Rate?	21
3.3.7	How is the Foam Applied?	21
3.3.8	Under What Conditions are the Fire Tests Performed?	22
3.3.9	What Burnback Test is Used?	22
3.3.10	When are the Fire Tests Carried Out?	22
3.4	Periodic Testing of Foam Concentrates	22
3.4.1	General	22
3.4.2	Collection of Foam Concentrate Samples	23
3.4.3	Typical Physical Property Tests	24
	(a) Specific Gravity (Relative Density)	24
	(b) pH (Acidity/Alkalinity)	24
	(c) Sediment (Sludge)	24
	(d) Spreading Coefficient	24
3.4.4	Periodic Fire Tests	24

Chapter 4 The Properties of Finished Foams and The Effects of These on Firefighting Performance   27

4.1	General	27
4.2	Working	27
4.3	Foam Flow/Fluidity	28
4.4	Film Formation	28
4.5	Fuel Tolerance	29
4.6	Edge Sealing	30
4.7	Foam Blanket Stability/Drainage Time	30
4.8	Vapour Suppression	31
4.9	Burnback Resistance	31
4.10	Water-miscible Fuel Compatibility	32
4.11	Suitability For Subsurface (Base) Injection	32
4.12	Quality of Finished Foam	32
4.13	Compatibility of Finished Foams	33
4.13.1	With Other Finished Foams	33
4.13.2	With Dry Powder	33

4.14	Typical Characteristics of Finished Foams	33
4.14.1	General	33
4.14.2	Individual Foam Characteristics	34
	(a) P	34
	(b) FP	35
	(c) FFFP	35
	(d) Synthetic (SYNDET)	35
	(e) AFFF	36
	(f) Alcohol Resistant Foam Concentrates (AFFF-AR and FFFP-AR)	36
4.15	Environmental Impact of Firefighting Foams	37
4.15.1	General	37
4.15.2	Toxicity	37
4.15.3	Biodegradability	38

Chapter 5 Equipment   39

5.1	General	39
5.2	Foam-Making Equipment	39
5.2.1	General	39
5.2.2	LX Hand-held Foam-making Branches	40
	(a) How They Work	40
	(b) LX Foam-making Branch Performance	41
5.2.3	LX Hand-held Hosereel Foam Unit	42
5.2.4	LX Foam Generators	43
5.2.5	LX Foam Monitors	44
5.2.6	MX Hand-held Foam-making Branches	45
5.2.7	LX and MX Hand-held Water Branch 'Snap-on' Attachments	46
5.2.8	MX Foam Pourers	46
5.2.9	HX Foam Generators	46
5.3	Foam Concentrate Induction and Injection Equipment	50
5.3.1	General	50
5.3.2	In-line inductors	51
5.3.3	Round-the-pump Proportioners	52
5.3.4	Pressure Control Valves	56
5.3.5	Pressurised Foam Supply	56
	(a) General	56
	(b) Distribution Manifold	60
	(c) Metering Devices	60
	(d) Inline Foam Injection (Pelton Wheel)	61
	(e) Pre-induction Units	61
	(f) Direct Coupled Water Pump	62
5.3.6	Hosereel Foam Induction and Injection Systems	63
	(a) General	63
	(b) Premix	63
	(c) Round-the-pump	64
	(d) Injection in to Pump Inlet	64
	(e) In-line Inductors	64
	(f) Suggestions for an Operational Requirement for a Hosereel Induction System	64
5.4	Compressed Air Foam Systems (CAFS)	65
5.5	Methods For Checking Foam Solution Concentration as Produced by Foam-making Equipment	65

5.5.1	General	65
5.5.2	Refractometer Method	66
5.5.3	Flow Method	66

Chapter 6 Categories of Fire and the Use of Firefighting Foams Against Them — 69

6.1	Classes of Fire	69
6.1.1	Class A fires	69
6.1.2	Class B Fires	70
	(a) General	70
	(b) High Flash Point Water-immiscible Class B Liquids	71
	(c) Low Flash Point Water-immiscible Class B Liquids	71
	(d) Water-miscible Class B Liquids	72
6.1.3	Class C Fires	72
6.1.4	Class D Fires	72
6.2	Electrical Fires	72
6.3	Types of Liquid Fuel Fire	72
6.3.1	General	72
6.3.2	Spill Fires	73
6.3.3	Pool Fires	73
6.3.4	Spreading Fires	73
6.3.5	Running Fires	74
6.3.6	Other Terms	74

Chapter 7 Application Rates — 75

7.1	General	75
7.2	Critical Application Rate	75
7.3	Recommended Minimum Application Rate	75
7.3.1	General	75
7.3.2	Fires Involving Water-immiscible Class B Liquids	75
7.3.3	Fires Involving Water-miscible Class B Liquids	77
7.4	Optimum Application Rate	77
7.5	Overkill Rate	77
7.6	Continued Application Rate	78

References — 79

Further Reading — 80

Glossary of Terms – Firefighting Foams — 81

Firefighting Foam –
Technical

Firefighting Foam – Technical

Chapter 1 – Introduction

1.1 General

Firefighting foams have been developed primarily to deal with the hazards posed by liquid fuel fires.

Water is used for most firefighting incidents. However, it is generally ineffective against fires involving flammable liquids. This is because water has a density that is greater than most flammable liquids so, when applied, it quickly sinks below their surfaces, often without having any significant effect on the fire. However, when some burning liquids, such as heavy fuel oils and crude oils, become extremely hot, any water that is applied will begin to boil. The resulting rapid expansion as the water converts to steam may cause burning fuel to overflow its containment and the fire to spread – this event is known as a slop-over. Also, the water that sinks below the fuel will collect in the container and, should the container become full, this will result in the fuel overflowing.

Finished firefighting foams, on the other hand, consist of bubbles that are produced from a combination of a solution of firefighting foam concentrate and water that has then been mixed with air. These air filled bubbles form a blanket that floats on the surface of flammable liquids. In so doing, the foam suffocates the fire and can lead to the knockdown and extinction of the flames.

The low density of firefighting foam blankets also makes them useful for suppressing the release of vapour from flammable and other liquids. Special foam concentrates are available which allow vapour suppression of many toxic chemicals.

Water-miscible liquids, such as some polar solvents, can pose additional problems for firefighters. These quickly attack finished foams by extracting the water they contain. This rapidly leads to the complete destruction of the foam blanket. Consequently, special firefighting foams, generally known as 'alcohol resistant' foam concentrates, have been developed to deal with these particular types of liquid.

Some firefighting foams have also been developed specifically for use against class A fires.

The main properties of firefighting foams include:

- Expansion: the amount of finished foam produced from a foam solution when it is passed through foam-making equipment.

- Stability: the ability of the finished foam to retain its liquid content and to maintain the number, size and shape of its bubbles. In other words, its ability to remain intact.

- Fluidity: the ability of the finished foam to be projected on to, and to flow across, the liquid to be extinguished and/or protected.

- Contamination resistance: the ability of the finished foam to resist contamination by the liquid to which it is applied.

- Sealing and resealing: the ability of the foam blanket to reseal should breaks occur and its ability to seal against hot and irregular shaped objects.

- Knockdown and extinction: the ability of the finished foam to control and extinguish fires.

- Burn-back resistance: the ability of the finished foam, once formed on the fuel, to stay intact when subjected to heat and/or flame.

The performance of firefighting foams can be greatly influenced by:

- The type of foam-making equipment used and the way it is operated and maintained.
- The type of foam concentrate used.
- The type of fire and the fuel involved.
- The tactics of foam application.
- The rate at which the foam is applied.
- The quality of the water used.
- The length of pre-burn.

The most effective and efficient use of firefighting foam can only be achieved after full consideration has been given to all of the above factors.

This Volume of the Manual describes all aspects of firefighting foam and discusses the types of equipment typically used by the fire service to produce it. Topics covered include the properties of foam concentrates, finished foams and foam equipment; application rates; and the classes of, and types of, fire for which foam can be used.

Volume 2 of the Manual describes the operational use of foam including recommended minimum application rates and application techniques; practical scenario considerations; and the logistics involved in dealing with fires in storage tanks.

At the rear of this Volume, there is a glossary of terms used in this Manual and other terms that may be used in connection with firefighting foams.

It must be stressed that this Manual only gives general information on the use of firefighting foams. Incidents requiring the use of foam are varied and preplanning in support of an effective risk assessment at the commencement of an incident is of the utmost importance to ensure that the correct foams, equipment and tactics are selected and employed.

1.2 Historical Development of Firefighting Foams

1877 – Chemical foam, first patented by a British scientist.

1904 – First successful use of chemical foam. Used to extinguish an 11 metre diameter naphtha storage tank fire in Russia. Foam produced from mixing together large quantities of two chemical solutions.

1914 – Austrian engineers produce foam by introducing a powder into running water.

1920s – Protein foam concentrate first produced along with equipment designed for the production and delivery of this first 'mechanical' foam.

1930s – Development of early chemical foams with alcohol resisting properties. The concepts of aspiration and proportioning were developed for mechanical foam systems much as we know them today. Experimental work started on synthetic types of foam concentrate.

1940s – 3% Protein foam concentrates developed to offer space and weight savings over the existing 6% concentrates.

1950s – Low, medium and high expansion foams could now be produced from a single synthetic foam concentrate. First water-miscible liquid resistant mechanical foam concentrate developed.

1960s – Fluoroprotein and AFFF (Aqueous Film-forming Foam) foam concentrates developed. Improved alcohol resistant foams developed.

1970s – Further development of alcohol resistant foam concentrates to produce multi-purpose foams for use at 3% on hydrocarbons and 6% on water-miscible liquids. "Hazmat" foams developed for the suppression of vapour from hazardous materials.

1980s – Development of alcohol resistant foam concentrates to produce AFFF-AR (alcohol resistant AFFF). Development of fluoroprotein foams to produce FFFP (Film-forming Fluoroprotein) and multi-purpose FFFP-AR (Alcohol Resistant FFFP) foam concentrates.

1990s – Development of alcohol resistant foam concentrates to produce versions that can be used at 3% concentration on both hydrocarbons and water-miscible liquids. Introduction of class A foam concentrates.

1.3 How Foams Extinguish Fires

Firefighting foam is much lighter (less dense) than all liquid fuels and so it floats on their surfaces. The foam blankets that are formed help to knockdown and extinguish these fires in the following ways:

- By excluding air (oxygen) from the fuel surface.

- By separating the flames from the fuel surface.

- By restricting the release of flammable vapour from the surface of the fuel.

- By forming a radiant heat barrier which can help to reduce heat feedback from flames to the fuel and hence reduce the production of flammable vapour.

- By cooling the fuel surface and any metal surfaces as the foam solution drains out of the foam blanket. This process also produces steam which dilutes the oxygen around the fire.

1.4 Production of Finished Foam

1.4.1 General

Finished foam is produced from three main ingredients; foam concentrate, water and air. There are usually two stages in its production. The first stage is to mix foam concentrate with water to produce a foam solution. The foam concentrate must be mixed into the water in the correct proportions (usually expressed as a percentage) in order to ensure optimum foam production and firefighting performance. This proportioning is normally carried out by the use of inductors (or proportioners) or other similar equipment. This results in the production of a 'premix' foam solution. In other words, the foam concentrate and water have been mixed together prior to arriving at the foam-making equipment. Occasionally, premix solutions are produced by mixing the correct proportions of water and foam concentrate in a container, such as an appliance tank, prior to pumping to the foam-making equipment. In addition, some types of foam-making equipment are fitted with a means of picking up foam concentrate at the equipment; these are known as 'self-inducing' with the mixing taking place in the foam-making equipment itself.

The second stage is the addition of air to the foam solution to make bubbles (aspiration) to produce the finished foam. The amount of air added depends on the type of equipment used. Hand-held foam-making branches generally only mix relatively small amounts of air into the foam solution. Consequently, these produce finished foam with low expansion (LX) ratios, that is to say, the ratio of the volume of the finished foam produced by the nozzle, to the volume of the foam solution used to produce it, is 20:1 or less. Other equipment is available which can produce medium expansion foam (MX) with expansion ratios of more than 20:1 but less than 200:1, and high expansion foam (HX) with expansion ratios of more than 200:1 and possibly in excess of 1000:1.

The following Sections describe in more detail some of the important factors of foam production that were introduced above.

1.4.2 Percentage Concentration

All foams are usually supplied as liquid concentrates. These must be mixed with water, to form a foam solution, before they can be applied to fires. They are generally supplied by manufacturers as either 6%, 3% or 1% foam concentrates. These have been designed to be mixed with water as follows:

- **6% concentrates**
  6 parts foam concentrate in 94 parts water,

- **3% concentrates**
  3 parts foam concentrate in 97 parts water,

- **1% concentrates**
  1 part foam concentrate in 99 parts water.

1% concentrate is basically six times as strong as 6% concentrate, and 3% concentrate is twice as strong as 6% concentrate. However, the firefighting characteristics of finished foam produced from 1%, 3% and 6% concentrates of a particular type of manufacturer's foam should be virtually identical.

The lower the percentage concentration, the less foam concentrate that is required to make finished foam. The use of say 3% foam concentrate instead of 6% foam concentrate can result in a halving of the amount of storage space required for the foam concentrate, with similar reductions in weight and transportation costs, while maintaining the same firefighting capability. Not all foam concentrates are available in the highly concentrated 1% form, e.g. alcohol resistant and protein based foam concentrates. This is because there are technical limits to the maximum usage concentrations of some of the constituents of foam concentrates.

It is extremely important that the foam induction equipment used is set to the correct percentage. If 3% concentrate is induced by an induction system set for 6% concentrate, then twice the correct amount of foam concentrate will be used creating a foam solution rich in foam concentrate. Not only will this result in the foam supply being depleted very quickly and an expensive waste of foam concentrate, but it will also lead to finished foam with less than optimum firefighting performance, mainly due to the foam being too stiff to flow adequately. Alternatively, using 3% foam concentrate where the system is set for 1% will result in a solution with too little concentrate to make foam with adequate firefighting performance.

It is also very important to have compatibility of foam-making equipment and induction equipment, and just as importantly, foam induction equipment must be checked regularly to ensure that it is operating correctly and giving an accurate rate of induction.

1.4.3 Aspiration

Once the correctly mixed foam solution has been delivered to the end of a hose line, there are a number of forms in which it can be applied to the fire. Generally, foam application is referred to as being either 'aspirated' or 'non-aspirated':

- **'Aspirated'** foam is made when the foam solution is passed through purpose designed foam-making equipment, such as a foam-making branch. These mix in air (aspirate) and then agitate the mixture sufficiently to produce uniformly sized bubbles (finished foam).

- **'Non-aspirated'** implies that no aspiration of the foam solution has taken place.

Consequently, the term 'non-aspirated foam' is often used incorrectly to describe the product of a foam solution that has been passed through equipment that has not been specifically designed to produce foam, such as a water branch. However, the use of this type of equipment will often result in some aspiration of a foam solution. This is because air is usually entrained into the jet or spray of foam solution:

- As it leaves the branch.

- As it travels through the air due to the turbulence produced by the stream.

- When it strikes an object. This causes further turbulence and air mixing.

There is sufficient air entrained by these processes to produce a foam of very low expansion (often with an expansion ratio of less than 4:1).

To more accurately describe the different types of finished foam produced, the terms 'primary' or 'secondary' aspirated are preferred:

- **Primary aspirated foam** – finished foam that is produced by purpose designed foam-making equipment.

- **Secondary aspirated foam** – finished foam that is produced by all other means, usually standard water devices.

Secondary aspiration will normally result in a poor quality foam being produced, due to insufficient agitation of the foam/air mixture. That is to say, the foam will generally have a very low expansion ratio and a very short drainage time (see below). However, foam blankets with short drainage times can be advantageous if rapid film-formation on a fuel is required (see Chapter 4, Section 4.4).

It is highly unlikely that a foam solution can be applied operationally to a fire in such a way that no aspiration occurs. However, should such circumstances occur, then this would be referred to as a non-aspirated application. Some water additives, such as wetting agents, may be formulated so that they do not foam; use of these types of additive would result in non-aspirated application, even through purpose designed foam-making equipment.

1.5 Foam Expansion Ratios

1.5.1 General

As mentioned previously, finished foam is usually classified as being either low, medium or high expansion. The expansion, or more strictly the expansion ratio, of a foam is the ratio of the volume of the finished foam to the volume of the foam solution used to produce it. For example, if 100 litres of foam solution were passed through a foam-making branch and 800 litres of foam were produced, then the expansion ratio of the foam would be calculated as follows:

Expansion ratio

= volume of foam : volume of foam solution

= 800 litres : 100 litres

= 8 : 1

This foam would also be referred to as having an expansion of 8.

Typical firefighting foam expansion ratio ranges are:

- **Low expansion** less than or equal to 20:1
- **Medium expansion** greater than 20:1 but less than or equal to 200:1
- **High expansion** greater than 200:1

Secondary aspirated foams generally have an expansion ratio of less than 4:1.

1.5.2 Equipment Used For Generating Different Expansion Ratio Foams

Primary aspirated low expansion foams are usually produced by using purpose designed foam-making branches or mechanical generators.

Secondary aspirated low expansion foams are usually produced by using standard water delivery devices although some purpose designed large capacity monitors have been produced for this particular type of application (see Volume 2).

Medium and high expansion foams are usually primary aspirated through special foam-making equipment. This equipment produces foam by spraying the foam solution on to a mesh screen or net. Air is then blown through the net or mesh either by entrainment caused by the spray nozzle, or by an hydraulic, electric or petrol motor driven fan.

1.5.3 Foam Concentrates

The amount that a foam solution can be aspirated not only depends on the equipment, but also on the foam concentrate that is used. For instance, synthetic detergent (SYNDET) foam concentrates are the only type that can be used to produce low, medium and high expansion foams; protein foam concentrates can only be used to produce low expansion foam and the remaining commonly used foam concentrates (i.e. AFFF, AFFF-AR, FP, FFFP and FFFP-AR, see Chapter 2) are mostly intended for use at low expansion, although they can also be used to produce medium expansion foam.

For flammable liquid fuel fires, effective secondary aspirated foam can only be produced using a film-forming foam concentrate.

Chapters 2, 3 and 4 discuss in detail the various types and properties of foam concentrates and finished foams.

1.5.4 Typical Uses and Properties of Low, Medium and High Expansion Finished Foams

The various expansion ratios are typically used for the following applications:

- **Primary Aspirated Finished Foams**

Low expansion
    Large flammable liquid fires (i.e. storage tanks, tank bunds)
    Road traffic accidents
    Flammable liquid spill fires
    Vapour suppression
    Helidecks
    Jetties
    Aircraft crash rescue
    Portable fire extinguishers

Medium expansion
    Vapour suppression
    Flammable liquid storage tank bunds
    Small cable ducts
    Small fires involving flammable liquids, such as those following road traffic accidents
    Transformer protection

High expansion
    Knockdown and extinction in, and protection of, large volumes such as warehouses, aircraft hangars, cellars, ships' holds, mine shafts, etc.
    Large cable ducts
    Vapour suppression (including cryogenic liquids such as LNG/LPG)

- **Secondary Aspirated Finished Foams**

Large flammable liquid fires (i.e. storage tanks, tank bunds)
Helidecks
Aircraft crash rescue
Portable fire extinguishers

Low expansion finished foams **can be projected over reasonably long distances and heights** making them suitable in many situations for use against fires in large storage tanks.

Medium expansion finished foam **can only be projected over small distances**. However, with expansions of between 20 and 200, large quantities of foam are produced from relatively small quantities of foam solution. This, combined with its ability to flow relatively easily, makes medium expansion foam ideal for covering large areas quickly.

High expansion finished foam flows directly out of the foam-making equipment and **is not projected any appreciable distance**. Its coverage of large areas can also be slow but the immense quantity of foam produced (expansion ratios are sometimes in excess of 1000:1) can quickly fill large enclosures. Often, flexible ducting is required to transport the foam to the fire. Due to its volume and lightness, high expansion foam is more likely than low and medium expansion foam, to break up in moderately strong wind conditions (Reference 1).

The equipment used to produce secondary aspirated foam is often standard water type branches and nozzles although there are some specifically designed large capacity nozzles available. The foam produced in this way is not well worked (see Chapter 4, Section 4.2), has a very low expansion ratio and short drainage time, and tends to be very fluid. These properties, combined with the film-forming nature of the foam concentrates used, can result in a finished foam blanket that can quickly knockdown and extinguish fires of **some** liquid hydrocarbon liquid fuels. This ability can make them ideal for use in certain firefighting situations such as aircraft crash rescue. However, the foam blanket tends to collapse quickly, so providing very poor security and resistance to burnback.

Secondary aspirated foam can be thrown over a greater distance than is possible with primary aspirated low expansion foam. This has resulted in equipment being designed specifically to project secondary aspirated foam into large storage tank fires. Manufacturers of this equipment recommend the use of film-forming foam concentrate types for such applications. They claim that the finished foam produced usually has an expansion ratio of less than 4:1.

Firefighting Foam – Technical

Chapter 2 – Foam Concentrates

2.1 Types of Foam Concentrate

2.1.1 General

There are a number of different types of foam concentrate available. Each type normally falls into one of the two main foam concentrate groups, that is to say, they are either protein based or synthetic based, depending on the chemicals used to produce them.

- **Protein based** foam concentrates include:
    Protein (P)
    Fluoroprotein (FP)
    Film-forming fluoroprotein (FFFP)
    Alcohol resistant FFFP (FFFP-AR)

- **Synthetic based** foam concentrates include:
    Synthetic detergent (SYNDET)
    Aqueous film-forming foam (AFFF)
    Alcohol resistant AFFF (AFFF-AR)

The characteristics of each of these foam concentrates, and the finished foams produced from them, varies. As a result, each of them has particular properties that makes them suitable for some applications and unsuitable for others.

For protein based foam concentrates, the basic chemical constituent is hydrolysed protein, obtained from natural animal or vegetable sources. It is the hydrolysed protein (the 'foaming agent') that enables bubbles to be produced.

For synthetic based foam concentrates, the basic constituents are detergent based foaming agents.

To enhance the firefighting properties of these basic constituents, and hence produce the different foam concentrate types, chemicals are added.

Various types of surface active agents (or surfactants) are added to many firefighting foam concentrates. These are used to reduce the amount of fuel picked up by the finished foam on impact with fuel (i.e. they increase fuel tolerance) and to increase the fluidity of the finished foam (i.e. they make it easier for finished foam to flow over some fuels and other surfaces).

Surface active agents are also used as foaming agents because they readily produce foam bubbles when mixed with water. Consequently, hydrocarbon surface active agents, or as they are more commonly known, synthetic detergents, are the main constituents of synthetic based foam concentrates. Surface active agents are also used in some protein based foams.

Surface active agents can help to reduce the surface tension of water. This not only helps in the formation of foam bubbles but also increases the ability of the water to penetrate and spread. This is particularly important when fighting class A fires because it can help water to penetrate and cool the burning material.

In film-forming foam concentrates, surface active agents form an aqueous film of foam solution which, in certain conditions, can rapidly spread over the surface of some burning hydrocarbons to aid knockdown and extinction.

Other chemicals may also be added to foam concentrates. These include corrosion inhibitors, solvents (to reduce viscosity and to enhance foaming properties), preservatives (to prevent the growth of bacteria and moulds), stabilisers (to help maintain foam bubble stability) and anti-freeze chemicals. These all help to prevent various problems that could arise if only the basic chemical constituents of the foam concentrates were used.

In addition to the two main foam concentrate groups, other specialised foam concentrates and water additives are available, in particular:

- **Hazmat foam concentrates** – for vapour suppression of toxic, odorous and/or flammable materials.
- **Wetting agents** for increasing the penetrating abilities of water.
- **Class A foam concentrates** – primarily for use on class A fires.
- **Fuel emulsifiers** – emulsion forming additives for use primarily on class B fires for firefighting and to prevent re-ignition.

Note that P, FP, FFFP, SYNDET and AFFF concentrates are often referred to as 'conventional' foam concentrates in order to distinguish them from alcohol resistant foam concentrates and the specialised foam concentrates and water additives mentioned above.

There are many companies manufacturing foam concentrates and the quality of the products varies from manufacturer to manufacturer. In addition, the quality of a particular manufacturer's version of a foam concentrate may vary slightly on a daily basis due to acceptable variations in the base materials used and other factors involved in the manufacturing process. To complicate this even further, some manufacturers produce different grades of the same foam concentrate type for different markets and, obviously, for use at different concentrations.

Consequently, the information contained within this Chapter gives an indication of the typical characteristics of each of the main types of foam concentrate. Good quality foam concentrates may have better characteristics, those of poor quality foam concentrates may be considerably worse. These characteristics will, in any case, vary depending on the equipment and tactics used, the size and type of incident and the fuel involved.

Some foam concentrate standards can help to distinguish good from bad quality products for certain applications. However, these standards need to be closely scrutinised to ensure that they meet the wide range of fire service requirements (see Chapter 3).

This Chapter provides information on each of the different types of foam concentrate and water additives mentioned above. Information is also given on the storage and handling characteristics of foam concentrates. However, the manufacturers should always be consulted regarding the suitability of materials used for the storage and handling of their products.

2.1.2 Protein Based Foam Concentrates

(a) Protein (P)

Protein foam concentrates are liquids that contain hydrolysed protein with, typically, the addition of stabilising additives and inhibitors to help prevent corrosion, resist bacterial decomposition, control viscosity and improve their shelf life. Chemical additives can include salts of iron and calcium, sodium chloride and solvent.

The starting materials for production, which provide the protein base product, include: soya beans, corn gluten, animal blood, horn and hoof meal, waste fish products and feather meal.

Protein foam concentrates are inexpensive and are usually manufactured for use at 3% or 6% concentrations. Versions are available that can be mixed with sea and fresh water. They are only intended for the production of low expansion finished foams.

In the past, protein foam concentrates have been widely used by industry, the fire service, the armed forces and aviation authorities throughout the world. They have now been largely superseded by fluoroproteins and film-forming foam concentrates although large stocks are still sometimes held.

Often, protein foam concentrates do not contain corrosion inhibitors as the concentrate is not considered to be particularly corrosive. However, incidents have indicated that some corrosion has taken place in unprotected carbon steel bulk storage containers. Consequently, materials such as epoxy coated carbon steel, GRP (Glass Reinforced Plastic) and polyethylene should be considered for the storage of protein foam concentrate.

(b) Fluoroprotein (FP)

FP foam concentrates basically consist of protein foam concentrates with the addition of fluorinated surface active agents (fluorosurfactants). The addition of fluorosurfactants provides oleophobic (oil repellent) properties and makes the finished foam more fluid. This greatly improves the fire knockdown performance of the finished foam when compared to that of protein foam. Other additives can include solvent, sodium chloride, iron, magnesium and zinc.

FP foam concentrates are usually available for use at 3% or 6% concentrations and versions are available for use with sea and fresh water. They are only marginally more expensive than protein foam concentrates.

FP foam concentrates are primarily intended for the production of low expansion foams although they have also proved effective when used to produce medium expansion foam. They are not recommended for the production of high expansion foam.

Uses are widespread in the fire service, the petrochemical industry and armed forces throughout the world. As with protein foam concentrates, corrosion inhibitors are not often included. However, consideration should be given to constructing bulk storage containers from materials such as epoxy coated carbon steel, GRP or polyethylene.

(c) Film-forming Fluoroprotein (FFFP)

FFFP foam concentrates are based on FP foam concentrates with the addition of film-forming fluorinated surface active agents. Under certain conditions, this combination of chemicals can, as well as producing a foam blanket, allow a very thin vapour sealing film of foam solution to spread over the surface of **some** liquid hydrocarbons.

FFFP foam concentrates are usually available for use at 3% or 6% concentrations. They are primarily intended for the production of low expansion foam although they can also be used to produce medium expansion foam. Also, due to their film-forming properties, they can be applied secondary aspirated and can be used to tackle class A fires.

FFFP foam concentrates are not recommended for the production of high expansion foam.

FFFP foam concentrates are more expensive than P and FP foam concentrates.

As with P and FP foam concentrates, consideration should be given to constructing bulk storage containers from materials such as epoxy coated carbon steel, GRP or polyethylene.

2.1.3 Synthetic Based Foam Concentrates

(a) Synthetic Detergent (SYNDET)

SYNDET foam concentrates were developed from early synthetic detergent foams and are based on a mixture of anionic hydrocarbon surface active agents, solvents and foam stabilisers.

SYNDET foam concentrates are versatile, as they can be used to produce low, medium and high expansion foams. They can also be used on class A and class B fires. In the UK, their use is usually limited to medium and high expansion foams. However, in other European countries such as Germany and Sweden, SYNDET foam is used for low expansion applications.

SYNDET foam concentrates are usually manufactured for use at between 1% and 3% concentrations and versions are available for use with sea and fresh water. They are of similar cost to P and FP foam concentrates.

Manufacturers have indicated that SYNDET foam concentrates are not particularly corrosive. However, testing (Reference 2) and reports received from brigades indicate that adverse corrosion and degradation effects can occur with materials such as epoxy coated carbon steel, GRP and aluminium. Materials that should be considered for bulk storage containers and equipment for both concentrate and solution are 316 stainless steel or polyethylene.

(b) Aqueous Film-forming Foam (AFFF)

AFFF foam concentrates are solutions of fluorocarbon surface active agents and synthetic foaming agents. Under certain conditions, this combination of chemicals can, as well as producing a foam blanket, allow a very thin vapour sealing film of foam solution to spread over the surface of some liquid hydrocarbons.

AFFF foam concentrates are usually available for use at 1%, 3% or 6% concentrations and versions are available for use with fresh and sea water. They are primarily intended for the production of low expansion foams although they can also be used to produce medium expansion foams. Due to their film-forming properties, they can be applied secondary aspirated and can be used to tackle class A fires. AFFF foam concentrates are not recommended for the production of high expansion foam.

AFFF foam concentrates are of similar cost to FFFP foam concentrates.

AFFF is widely accepted for crash rescue fire-fighting uses and on less volatile fuels such as kerosene and diesel oil. It is widely used offshore secondary aspirated for helideck protection at a concentration of 1%.

Problems have been experienced when attempting to extinguish fires involving liquids with high vapour pressures, such as hexane and high octane petrol, where quantities of vapour have penetrated thin, very low expansion (secondary aspirated) AFFF foam blankets.

AFFF foam concentrate is not particularly corrosive and contains no special corrosion inhibitors. However, its surface active agent content causes the concentrate to be more searching than water and therefore more corrosive. Materials that should be considered for materials for storage containers and handling equipment are stainless steel, GRP, epoxy lined carbon steel and polyethylene.

2.1.4 Alcohol Resistant Foam Concentrates (AFFF-AR and FFFP-AR)

Alcohol resistant foam concentrates have been developed to deal with fires involving water-miscible liquids such as alcohols and some petrol blends containing high levels of alcohols and other similar fuel performance improvers.

Two types of alcohol resistant foam concentrate are in general use in the UK fire service; those based on synthetic aqueous film-forming foams (AFFF-AR) and those based on film-forming fluoroprotein foams (FFFP-AR). Alcohol resistant foams can also usually be used on hydrocarbon fuels and because of this are sometimes known as multi-purpose foams.

Non-alcohol resistant foam concentrates (i.e. P, FP, FFFP, SYNDET and AFFF) are not suitable for use on water-miscible liquids because their finished foam blankets quickly disintegrate on contact with these liquids. This happens because the water contained in the foam rapidly mixes with, and is extracted by, the water-miscible liquids causing the foam to quickly break down and disappear.

AFFF-AR and FFFP-AR foam concentrates contain a polymeric additive which rapidly falls to the surface of a water-miscible liquid when the finished foam comes into contact with it. The polymeric additive forms a tough 'skin' (also known as a 'raft' or 'membrane') on the surface of the liquid. Once formed, the water-miscible liquid cannot penetrate this skin and is hence unable to attack the finished foam above it; conventional foams cannot form these water-miscible liquid resistant skins.

It should be noted that the polymeric membrane is not formed when alcohol resistant foams are applied to hydrocarbon fuels. It is also important to note that although AFFF-AR and FFFP-AR finished foams form aqueous films on **some** liquid hydrocarbon fuels, it is not possible for them, or any other foams, to form aqueous films on water-miscible liquids. They will, however, form an aqueous film between the polymeric membrane and the finished foam blanket. This may help to quicken the repair of any breaks that may occur in the polymeric layer.

Alcohol resistant foam concentrates are normally used at 6% concentration for application to fires of water-miscible fuels, such as most polar solvents, and at 3% concentration on liquid hydrocarbon fuel fires. However, some alcohol resistant foam concentrates have been specifically designed for use at 3% concentration on both water-miscible and hydrocarbon fuels. The

3%/6% concentrates are similar in price to standard AFFF and FFFP concentrates whereas the manufacturers tend to charge more for the single 3% concentrates.

The viscosity (see this Chapter, Section 2.2.2) of alcohol resistant foam concentrates can vary enormously; some flow relatively easily while it can be difficult to pour others out of their containers. In addition, they become more viscous with falling temperature. Consequently, if these foam concentrates are to be used, it is important to ensure that existing induction equipment will pick them up at the correct rate when using typical operational equipment and conditions. For instance, when using the more viscous foam concentrates, it is likely that in-line inductor dial settings will be incorrect and not as much concentrate as indicated will be picked-up. As a result, when using these viscous foam concentrates, foam induction systems may need to be re-calibrated. In addition, as the temperature of the alcohol resistant foam concentrates falls towards freezing (0°C), the rate at which they are picked up by the induction system will reduce further, due to increasing viscosity, possibly even making the re-calibration inaccurate.

Alcohol resistant foam concentrates are primarily designed for the production of low expansion foams although they may also be used to produce medium expansion foams for application to hydrocarbon and water-miscible liquids. Versions are available for use with sea and fresh water.

For AFFF-AR, the suggested materials for bulk storage containers and equipment are the same as AFFF, that is stainless steel, GRP, epoxy lined carbon steel and polyethylene.

For FFFP-AR, as with P, FP and FFFP foam concentrates, it is suggested that bulk storage containers should ideally be constructed from materials such as epoxy coated carbon steel, GRP or polyethylene.

Alcohol resistant versions of P, FP and SYNDET foam concentrates are available although they are uncommon in the UK. They are used in other European countries, in particular, FP-AR is widely used in France.

2.1.5 Hazmat Foam Concentrates

Many materials used in industrial and chemical processes release toxic, odorous and/or flammable vapour when in contact with the atmosphere. If a spill occurs, the hazard can be reduced by suppressing the released vapour until the spill can be neutralised and disposed of.

Some of the conventional firefighting foams discussed above may be used for vapour suppression on spills of flammable and combustible products. Also, a certain amount of success has been achieved with them on toxic spills. However, many chemicals destroy firefighting foams either by reacting with them or by extracting the water from foam blankets. Alcohol resistant foams can be effective on some toxic spills and flammable, combustible and water-miscible liquids.

Hazmat foam concentrates have been designed to be effective on products which destroy foams by chemically reacting with them. Versions of these foam concentrates are available that have been formulated to be resistant to either extreme acidity or extreme alkalinity. They are often used to produce medium expansion foams with optimum expansion ratios of around 60:1.

Developments in this area include an additive for use in conjunction with one particular alcohol resistant foam concentrate that significantly slows down the drainage rate of the finished foam to produce a very stable foam blanket that lasts in excess of 12 hours. This can be used on hazardous materials and is easily washed away with a water spray after use. However, additional equipment is required to mix the additive into the foam solution line on application.

If there is doubt concerning the suitability of a foam concentrate for a particular task, the manufacturer of the foam concentrate should be consulted to ensure that it can be used safely and successfully.

2.1.6 Wetting Agents

Wetting agents are liquids which, when added to water in the required proportion, reduce the surface tension of the water and increase its penetrating and spreading abilities. They may also provide emulsification (see Section 2.1.8 below) and foaming characteristics. Dedicated wetting agents are available although some manufacturers of film-forming and SYNDET foam concentrates state that these too may be used as wetting agents.

Dedicated wetting agents are generally used at concentrations of up to 1%. In addition, some film-forming and SYNDET foam concentrates intended for use at 3% on hydrocarbon fuel fires can be used as wetting agents at concentrations of between 0.5% and 3.0%. Wetting agents are generally recommended for use in either non-aspirated or secondary aspirated application through standard water branches.

Some dedicated wetting agents are also recommended for use on class B fires. Some limited tests (Reference 3) have indicated that they are unsuitable for this type of application.

2.1.7 Class A Foam Concentrates

The term 'Class A Foam' originated in the USA and is used to describe foam concentrates that are primarily intended for use on class A fires. They have been in use in the USA for more than 20 years in fighting wildland fires but more recently they have been gaining in acceptance there for use in structural firefighting.

Class A foam concentrates are often synthetic detergent foam concentrates that have been formulated for use on class A fires only. They are claimed to reduce the surface tension of water to increase its capacity to spread and penetrate class A fuels. Consequently, if this is the case, some class A foams may also be defined as wetting agents (see above). Generally, they are formulated for use at concentrations of up to 1%. They are mostly intended for use either non-aspirated or secondary aspirated using standard water branches. Some have also found use in compressed air foam systems (CAFS – see Chapter 5, Section 5.4).

Tests carried out in the UK (Reference 4) have shown that class A foams, and the two conventional foams tested (i.e. AFFF and SYNDET), perform no better than water when used to extinguish fires in wooden pallets.

2.1.8 Fuel Emulsifiers

Fuel emulsifiers are mixtures of emulsifiers, wetting agents and other additives. They are generally designed for use at concentrations between 0.5% and 6% in water and may or may not produce foams. They are formulated specifically for application to class B petroleum based fuels although some manufacturers also recommend their use for class A fires.

Fuel emulsifiers are oleophilic; in others words they are 'oil liking'. Consequently, on application to petroleum based fuels, it is claimed that the fuel emulsifier solution mixes with the fuel to form an emulsion which consists of fuel molecules encapsulated in water molecules. This is said by the manufacturers to significantly reduce the amount of vapour released by the fuel making the mixture incapable of sustaining combustion. When used against petroleum fuel fires, sufficient mixing of the emulsifier with the fuel, by very vigorous direct application to the surface of the fuel, is said to result in rapid knockdown and extinction of the fire. In addition it is claimed that because an emulsion has been formed and the fuel molecules have been encapsulated, re-ignition should not occur and that the mixture is then suitable for disposal with no risk of re-ignition.

On class A fires, fuel emulsifiers are claimed to simply act as class A foams (see above).

Emulsifiers have only recently been introduced and their performance relative to other foam concentrates and firefighting media has yet to be proven in the UK.

2.2 Handling and Storage of Foam Concentrates

2.2.1 Compatibility

Different types and makes of foam concentrate are not generally compatible and manufacturers' advice and recommendations should be followed. Mixing incompatible foam concentrates may cause sludge and sedimentation to form in the concentrates which may lead to blockages in induction systems and other equipment. Mixing of incompatible foam concentrates is also likely to lead to poor firefighting foam being produced with an associated reduction in firefighting performance.

Consequently, the ground rules to ensure that incompatible foam concentrates are not mixed together are as follows:

- Do not mix together different types, grades, brands, or concentrations of foam concentrate without first consulting the manufacturer(s). All possible adverse effects, such as reduced shelf life, formation of sludge, reduction in firefighting performance etc., should be explored with the manufacturer and understood. If the manufacturer(s) agree to this mixing, it is likely that the resulting foam concentrate mixture will tend to exhibit the least effective properties of each of the foam concentrates mixed.

- When changing over from one type of foam concentrate to another, especially in bulk storage or fire appliance tanks, first ensure that all of the old type has been removed, and the tank and equipment have been thoroughly cleaned and dried before refilling. Ensure that the new foam concentrate is compatible with the material of manufacture of the storage container.

- The chemical properties of foam concentrates can change with time and storage conditions. Consequently, even a new batch of the same brand and grade might cause difficulties when mixed with older stock especially if deterioration of the old stock has taken place. Manufacturers should be consulted if there are any doubts. Freeze protected and non-freeze protected versions of the same brand can be mixed but there will obviously be a reduction in the freeze-protection of the foam concentrates.

2.2.2 Viscosity

Viscosity is a measure of how well a liquid will flow. A low viscosity is often desirable because it improves the flow characteristics of a foam concentrate through pick-up tubes, pipework and induction equipment. Liquids are generally classed as either being non-Newtonian or Newtonian.

Many alcohol resistant foam concentrates are considered to be non-Newtonian pseudo-plastic liquids. For these liquids, as their flow increases, their viscosity decreases and so they flow more easily. Consequently, getting them to flow initially can be difficult, but once flowing, their viscosity reduces to a more acceptable level.

In contrast, the viscosity of Newtonian liquids, such as most non-alcohol resistant foam concentrates, remains the same no matter how quickly or slowly they are flowing.

Viscosity will also vary with foam concentrate type and with concentration. AFFF foam concentrates at 3% and 6% concentrations tend to be the least viscous, closely followed by P, FP and FFFP foam concentrates at 6%. AFFF at 1% and SYNDET foams, P, FP and FFFP foam concentrates at 3% concentration are appreciably more viscous than these. The alcohol resistant foams are often the most viscous although recent developments have dramatically reduced the viscosity of some products.

In addition, the viscosity of all foam concentrates will vary with temperature and may be affected by the age of the foam concentrate. Manufacturers often state the viscosity of their products when measured at 20°C; lower temperatures will result in much higher viscosity.

Manufacturers may also quote a 'Lowest Use Temperature' or 'Minimum Use Temperature' for their foam concentrates. The definition of these terms varies but they should be used to indicate the temperature below which foam concentrates cannot be used through induction systems. However, these figures must be treated with some caution because foam concentrates above these low temperatures may still have high viscosity which will prevent them being picked up at the correct rate by most foam concentrate induction systems.

Induction equipment should be checked for accuracy both when the foam concentrate is at the lowest temperature at which it expected to be used and at 'normal' operating temperatures. With some foam induction systems, the use of high viscosity foam concentrates and some non-Newtonian pseudo-plastic foam concentrates, will result in little, or no, foam concentrate being picked-up.

2.2.3 Corrosion

An initial indication of how corrosive a liquid may be can be made by looking at how acidic or alkaline it is. The measure used for this is pH which is on a scale of 1 to 14. If the pH of a liquid is lower than 7 then it is an acid; if it is higher than 7, it is an alkaline. A liquid with a pH of 7 is referred to as neutral, being neither acid nor alkaline; pure water has a pH of 7.

Acidic liquids are usually the most corrosive to metals and alloys, particularly those containing iron, such as carbon steel or cast iron. Strong alkaline liquids can attack aluminium and zinc.

Firefighting foam concentrates can contain a high percentage of water; in some the water content can be as much as 80%. Consequently, most foam concentrates are nearly neutral with pH values of between 6.5 and 9.0. The limits of pH of a particular foam concentrate are normally given by the manufacturer and are determined in laboratories by using pH meters.

In addition to the effects of pH, surface active agents can increase corrosion mainly due to their cleaning and penetrating properties, although other chemical actions can also take place.

Foam concentrate manufacturers should always be consulted on the best materials for use with their products. However, testing (Reference 2) has indicated that UPVC, 60/40 brass, 70/30 brass and stainless steel may be the best materials for use in storing the types of foam concentrate most often used by the UK fire service (i.e. AFFF, AFFF-AR, FFFP, FFFP-AR, P, FP and SYNDET). Zinc (for galvanising) was found to be unacceptable for the storage of the P, FP, FFFP and FFFP-AR foam concentrates but was acceptable for the AFFF types. Aluminium was found to be an excellent material for the storage of the AFFF type foam concentrates but unacceptable for any of the others. GRP and epoxy coated materials were found to be acceptable for all but the SYNDET foam concentrate which produced severe damage in both materials; in particular it caused the epoxy coating to peel away from the underlying steel.

The effects of corrosion will not only lead to the gradual, or sometimes rapid, destruction of the storage containers, but it may also lead to serious chemical effects on the foam concentrates themselves, possibly leading to poor foam production and firefighting performance.

The corrosion and chemical effects can take many forms but a particularly serious consequence can be the formation of particles and very viscous products (sludge) in the foam concentrate. These effects can lead to blockages and other serious problems with induction systems and other equipment.

2.2.4 Storage and Use Temperature Conditions

Many P, FP, AFFF and FFFP foam concentrates are freeze protected for low temperature storage and use. Some manufacturers state that some of their foam concentrates can be used when they are at temperatures as low as -29°C.

Some manufacturers produce both freeze protected and non-freeze protected versions of their foam concentrates. Care must be taken with the non-freeze protected versions as some of these should not be subjected to freezing and their minimum use temperature is often around 2°C.

As mentioned previously (see Section 2.2.2 above), foam concentrates generally become more viscous the cooler they become. Consequently, the minimum use temperature given by manufacturers for their foam concentrates is often based on their assessment of how the viscosity of their products will affect the induction rate. When used at, or near, their minimum use temperature, the viscosity of some foam concentrates will be so great that they will not be picked-up at the correct rate by some foam induction equipment.

Manufacturers recommend minimum and maximum storage temperatures for their foam concentrates. This can be a very wide temperature range, for instance, some freeze protected foam concentrates can be stored at between -29°C and 60°C.

Care should be taken to ensure that foam concentrates are not subjected to temperatures outside of the ranges specified by the manufacturers. Should this occur, especially over long periods of time, then it is likely to seriously impair the firefighting performance of the foam concentrates.

It should be noted that some foam concentrates have recommended maximum storage temperatures of 40°C. It is quite possible for temperatures of this order to be regularly reached in storage containers kept in direct sunlight.

Storage at constant low temperatures, in the order of 10°C, will help to extend the shelf life of foam concentrates.

When stored under the conditions recommended by the manufacturer, most foam concentrates should last at least 10 years and some should remain in good condition for considerably longer. The condition of stored foam concentrates should be checked on a regular basis (see Chapter 3, Section 3.4).

2.2.5 Order of Use

Wherever possible, foam concentrates should be used in the order in which they were manufactured/delivered. This will help to prevent prolonged storage of foam concentrates and unwanted effects such as sedimentation and sludge that may occur with age. Writing the delivery date on the containers is a simple way of keeping track of the age of the foam concentrates. Some manufacturers print the date of manufacture on the container labels.

2.2.6 Storage Containers and Bulk Storage

Manufacturers often advise that their products should be kept in original, sealed containers to help to maintain the concentrates in good condition. These are often 20 or 25 litre cans, 200 litre drums or 1000 litre containers.

If original containers are not used, then the advice is to ensure that the storage containers are kept full and sealed to prevent evaporation and oxidisation of the foam concentrate due to the chemical reaction of the concentrate with air.

The use of pressure/vacuum vents in storage tanks are also sometimes recommended in order to reduce these effects. Sealing oils can also be used to cover the surface of the foam concentrate although pressure/vacuum vents will still be required.

The materials used for the construction of the containers and associated fittings, pumps etc. should also be carefully considered to ensure that corrosion, and a possible reduction in firefighting performance, does not occur (see Section 2.2.3 above).

The positioning of storage containers should also be a major consideration to ensure that the foam concentrates are not subjected to temperatures beyond the storage limits recommended by the manufacturers (see Section 2.2.4 above).

Containers that are refilled before being completely emptied may cause foam quality and firefighting performance problems even if the same type and make of foam concentrate is used. The foam concentrates may be incompatible (see Section 2.2.1 above) and the mixing of different ages of foam concentrate may produce unwanted side effects, such as sedimentation and sludge. Ideally, containers should be completely emptied, cleaned and dried before they are refilled.

Methods of transporting the foam concentrate and/or their containers to the fireground and then distributing the foam concentrate to foam making equipment also need to be carefully considered. Fixed bulk storage containers will require adequately specified and sized pumps and/or outlets (especially for gravity fed systems) to ensure foam concentrate supplies are loaded into mobile units in the shortest possible time. Mobile units should also have adequately specified and sized pumps and outlets to ensure quick delivery of the foam concentrate when on the fireground. The materials of construction of the containers and associated fittings on the mobile units should also be chosen with the corrosive and other effects of foam concentrates in mind.

Firefighting Foam – Technical

Chapter 3 – Foam Concentrate Standards and Periodic Testing

3.1 General

Foam concentrates should be purchased that comply with standards that are relevant to their use by the fire service. They should also be tested periodically to ensure that they have not degraded (e.g. due to ageing, accidental dilution or contamination).

Manufacturers usually produce their foam concentrates to comply with one or more foam concentrate standards. The following foam standards are often quoted in manufacturers literature:

Standard		Title
BS EN 1568		– Fire Extinguishing Media – Foam Concentrates (British/European Standard)
	Part 1	– Specification for medium expansion foam concentrates for surface application to water-immiscible liquids
	Part 2	– Specification for high expansion foam concentrates for surface application to water-immiscible liquids
	Part 3	– Specification for low expansion foam concentrates for surface application to water-immiscible liquids
	Part 4	– Specification for low expansion foam concentrates for surface application to water-miscible liquids
ISO 7203:1995		– Fire Extinguishing Media – Foam Concentrates (International Standards Organisation)
	Part 1	– Specification for low expansion foam concentrates for top application to water-immiscible liquids
	Part 2	– Specification for medium and high expansion foam concentrates for top application to water-immiscible liquids
	Part 3	– Specification for low expansion foam concentrates for top application to water-miscible liquids
DEF STAN 42-40		– Foam Liquids, Fire Extinguishing (Concentrates, Foam, Fire Extinguishing) (UK, Ministry of Defence)
DEF STAN 42-41		– Foam Liquids, Fire Extinguishing (Concentrates, Alcohol Resistant Foam, Fire Extinguishing) (UK, Ministry of Defence)
ICAO/CAA CAP 168		– Licensing of Aerodromes, Chapter 8, Appendix 8E, Foam Performance Levels, Specifications and Test Procedures (UK, Civil Aviation Authority)
UL 162		– Foam Equipment and Liquid Concentrates (USA, Underwriters Laboratories)
MIL-F-24385		– Fire Extinguishing Agent, Aqueous Film-forming Foam (AFFF) Liquid Concentrate, For Fresh and Sea Water (USA, Military/Navy)

Each of these standards has been produced in order to ensure the quality of particular foam concentrates for particular purposes:

- The British, International and European standards have been produced for procurement of all types of foam concentrates which meet minimum performance requirements for general firefighting applications.

- DEF STAN 42-40 specifies requirements for foam concentrates for controlling and extinguishing hydrocarbon fires in aircraft, ships and vehicles, as well as for general purpose use. The standard covers P, FP, AFFF and FFFP foam concentrates.

- DEF STAN 42-41 specifies requirements for alcohol resistant foam concentrates, for controlling and extinguishing fires where solvents and products containing solvents are bulk stored. The standard covers AFFF-AR and FFFP-AR foam concentrates when used at 6% concentration.

- The International Civil Aviation Authority (ICAO) specify performance standards for foam concentrates in their document Airport Services Manual Part 1 9137-AN/898 which supports the requirements to be met by Airport Fire Services to be compliant with ICAO Annex 14, Volume one (Aerodrome Design and Operations). The UK Civil Aviation Authority (CAA) has adopted the ICAO foam standard in its guidance document Civil Air Publication 168 (Licensing of Aerodromes).

- UL162 covers foam producing equipment and liquid foam concentrates used for the production and discharge of firefighting low expansion foam. UL is unique in that it is the firefighting 'system' that is approved (including the foam-making branch) and not the foam concentrate as an individual item.

- The 'MIL-F spec' was designed by the US navy to assess the suitability of 3% and 6% aspirated AFFF firefighting foams for crash fire situations. US Navy typical applications of AFFF include incidents on the flight decks of aircraft carriers where a quick knockdown of shallow spill fires is required to assist air crew survivability.

When purchasing foam concentrates, it is important to have some background knowledge of these standards in order to decide whether the foam concentrates complying with them are likely to be suitable for fire service use. Ideally, the standards themselves should be obtained and evaluated.

It should be remembered that the methods and evaluation techniques used may vary considerably from standard to standard. As a result, it can be very difficult and unwise to compare results achieved by one foam concentrate when tested to one standard with those achieved by a second foam concentrate when tested to another standard. In addition, the results of standard (small-scale) fire tests cannot be relied upon to predict the firefighting performance of foam concentrates when used on large fires.

Generally, foam concentrate standards consist of two main areas of testing:

- Physical property tests

- Fire tests

In Sections 3.2 and 3.3 of this Chapter, physical property tests and fire tests are discussed in general terms.

Once the concentrate has been purchased, it should be stored and used as recommended by the manufacturer or supplier (see Chapter 2, Section 2.2). However, the foam concentrate will eventually deteriorate and so it is important that foam stocks are periodically tested to ensure that their performance remains acceptable. Section 3.4 of this Chapter discusses periodic testing including typical physical property and fire tests that might be performed and also provides information on the collecting of representative foam concentrate samples from storage containers.

3.2 Physical Property Tests of Foam Concentrates

3.2.1 General

Physical property tests often include laboratory measurements of parameters such as pH (acidity/alkalinity), viscosity, specific gravity, sediment and the effects of accelerated ageing. Standards generally contain well defined methods and equipment for the measurement of these properties. The results of these tests can be used to compare the properties of the foam concentrate with minimum/maximum requirement limits set within standards or with previously tested foam concentrates.

The data provided by these tests can be used by manufacturers as bench marks for checking the consistency of later manufactured batches of foam concentrates (quality control).

The measurements can also be used for comparison purposes in order to determine the condition of foam concentrates after long periods of storage (see this Chapter, Section 3.4).

Most physical property tests are relatively simple and inexpensive to perform. Consequently, manufacturers are more likely to carry out physical property tests than carry out fire tests as part of their quality control procedures. However, physical property tests do not provide any useful information regarding the firefighting performance of foam concentrates.

A wide range of physical property tests are carried out as part of standard approvals processes, the following physical property tests are most often included:

3.2.2 Specific Gravity (Relative Density)

Specific gravity (or relative density) is a measure of the ratio of the mass of a given volume of foam concentrate to the mass of an equal volume of water. This is normally measured with the temperature of the foam concentrate and water at 20°C. Specific gravity can be used to determine whether a foam concentrate has been diluted or over concentrated.

3.2.3 pH (acidity/alkalinity)

pH is a measurement of the acidity to alkalinity of a liquid on a scale of 1 to 14. A pH of 7 is neutral (e.g. pure water), a pH of 1 is very acidic, a pH of 14 is very alkaline. Measurements of pH help to give an indication of the corrosion potential of the liquids (Section 2.2.3).

3.2.4 Sediment (Sludge)

Sediment is a measure of the amount, as a percentage by volume, of undissolved solids contained in the foam concentrate. Sediment is also sometimes known as sludge. Excess sediment can result in blockages and other serious problems with induction systems and other equipment.

3.2.5 Spreading Coefficient

Film-forming foam concentrates are formulated to form an aqueous film on the surface of some hydrocarbon liquids. Spreading coefficient is a measure of this ability.

This is determined in a laboratory by measuring the surface tensions of a solution of the foam concentrate and a hydrocarbon liquid (normally cyclohexane). In addition, the interfacial tension is also determined, by measuring the surface tension where the foam solution (top) and the hydrocarbon liquid (bottom) meet. A calculation is then performed to determine the spreading coefficient of the foam solution. The calculation is as follows:

Spreading coefficient
= Surface tension of the foam solution
 minus Surface tension of the hydrocarbon liquid
 minus Interfacial tension

If the spreading coefficient is positive, the foam solution will form an aqueous film on that particular hydrocarbon liquid and the foam concentrate is deemed to be 'film-forming'. If the spreading coefficient is negative, an aqueous film will not be formed and the foam concentrate is not film-forming.

Note that although a solution of the foam concentrate may form a film on cyclohexane, or what ever hydrocarbon liquid used, this does not necessarily mean it will form a film on this or any other hydrocarbon liquid under operational conditions (see Chapter 4, Section 4.4).

3.2.6 Effects of Freeze/Thaw

Freeze/thaw tests are used to determine the effects on a sample of foam concentrate of several cycles of cooling it below its freezing point and then thawing it out. Some standards require a selection of physical property tests to be carried out after the freeze/thaw cycle. The results of these are then compared with measurements made before the tests; any variations must fall within certain limits. Other standards simply require observation of the sample for evidence of solids, crystals or sludge.

3.2.7 Accelerated Ageing

Accelerated ageing is intended to determine the effects on a foam concentrate of long term storage. The test usually involves storing a sample of the foam concentrate at a high temperature (e.g. 60°C) for an extended period of time (e.g. 7 days). The foam concentrate is then allowed to cool and the effects on the foam concentrate are measured, normally by comparing before and after physical property tests.

3.2.8 Viscosity

Viscosity is a measure of how well a liquid will flow (see Chapter 2, Section 2.2.2). Liquids are generally classed as either being non-Newtonian or Newtonian. A low viscosity is often desirable because it improves the flow characteristics of a foam concentrate through pick-up tubes, pipework and induction equipment. The viscosity of the foam concentrate is usually measured either at 20°C or at its minimum use temperature.

3.3 Foam Concentrate Standard Fire Tests

3.3.1 General

Standard fire tests, that is those fire test methods that are contained in various foam concentrate standards (e.g. British, European and International standards) are used to assess the firefighting performance of foam concentrates under closely controlled, but artificial, conditions. The results of these tests can be used to compare the performance of foam concentrates with minimum/maximum requirements within the standards or with previously tested foam concentrates. Typically, timings are recorded to 90% extinction, 99% extinction, complete extinction and 25% or 100% burnback.

The surface area of the test fires varies, but it is usually in the region of $0.25m^2$ to around $4.5m^2$. Small standard fire tests are used by some manufacturers for quality control purposes during production although fire tests are usually considered to be environmentally unfriendly, inconvenient, costly and time consuming to perform.

Results of standard fire tests cannot be used to predict the firefighting performance of foams operationally although they do at least indicate that the foams can put out fires. They can also be used to ensure that the firefighting performance of foam concentrates has not deteriorated due to age, corrosion, contamination etc. However, this requires that the same test method and equipment have been used previously on the foam concentrate in order to enable a valid comparison to be made.

It should also be noted that all of the standards referred to in this Chapter of the Manual are for primary aspirated foams only; there are currently no standards available for determining the suitability of foam concentrates for fire service secondary aspirated use.

When looking at the suitability of standard fire tests for particular fire service related applications, the following questions should be addressed:

3.3.2 Is the Fuel Commonly Encountered Operationally?

Petrol is the most likely fuel to be encountered operationally. Fuels such as avtur, avgas and heptane are not as volatile as petrol and are generally easier to extinguish. Avtur and avgas may be in regular use at airfields but are rarely encountered elsewhere. Heptane is unlikely to be encountered operationally and is not representative of any fuel that is.

3.3.3 Is the Fuel Reproducible?

Is the fire test fuel manufactured to a tight enough specification so that the burning characteristics of the fires are always similar? The specifications for military and aviation grades of avtur and avgas can be strict which enables them to be used as test fuels. Heptane is a very reproducible fuel and this is the main reason why it is used as the test fuel in many standards. Various well defined grades of Heptane are available and the exact grade required for a particular standard fire test is normally specified.

In Europe, petrol is produced to European standards that allow variations in formulation within fairly large margins. This allows petrol to be produced economically but provides a fuel whose burning properties and effects on foam can vary considerably. These variations make petrol unsuitable for use as a standard test fuel.

3.3.4 How Long is the Preburn?

Preburn times (i.e. the time from ignition of the fuel until the application of foam) can vary from standard to standard. Short preburns are unlikely to allow the fuel burning rate and heat output to stabilise and will not allow the tray sides enough time to become hot. Longer preburns are more realistic and consequently the fires are likely to be more difficult to extinguish. Preburns of around a minute are often used for hydrocarbon fuels. This is a compromise between fuel costs and fire severity. Fires involving water-miscible fuels take much longer to stabilise and so the longer the preburn the better.

3.3.5 How Deep is the Fuel?

For hydrocarbon fires, the fuel depth should be at least 25mm (a spill fire – see Chapter 6, Section 6.3.2) or, preferably, deeper. This is likely to be a more realistic condition for the tests and will provide enough fuel for a reasonable preburn time and burnback test. However, it must be remembered that with an average hydrocarbon burning rate of 4mm per minute, a 25mm depth of a typical hydrocarbon fuel will only burn for around 6 minutes.
Fire tests involving water-miscible fuels should have a much greater depth. This is because their extinction can be aided by the dilution of the fuel with the applied foam solution.

Most standard fire tests involving hydrocarbons require there to be a depth of water (a water base) in the tray. This helps to ensure a consistent depth of fuel over the whole area of the tray and helps to prevent heat damage to the fabric of the fire tray. Fire tests involving water-miscible type fuels must not have water bases because these will dilute the fuel making it easier to extinguish.

3.3.6 What is the Application Rate?

The application rate should be above the critical application rate (see Chapter 7, Section 7.2) but should not be too high. If a high application rate is used then it is likely that the fire will be extinguished very easily, even with poor quality foam concentrates. The application rate should certainly not be any higher than the minimum recommended application rate for spill fires given in this Manual (see Chapter 7, Section 7.3).

3.3.7 How is the Foam Applied?

Some standards involve applying foam gently via a back-plate. Although it is recommended that foam should be applied gently when used operationally, this is not often possible. The better standards for foam concentrates for fire service use are those which require the foam to be applied forcefully to the surface of a burning fuel, i.e. the 'worst case' situation. Forceful application is far more testing of the firefighting capabilities of the foam, particularly its fuel tolerance.

Some standards specify that the foam-making branch should be in a fixed position, others allow it to be hand-held. Fixed branches are more likely to result in a repeatable fire test while the hand-held branch is more realistic. However, hand-held applications can result in variations in firefighting performance that can be attributed more to the operators experience and tactics than the properties of the foam alone. Fixed branches have the disadvantage that fire tests involving them will tend to favour the more fluid foams.

Generally, the test equipment used during standard fire tests makes finished foams that have lower expansion ratios and much longer drainage times. Consequently, the foams produced are not realistic because they are more stable and better worked (see Chapter 4, Section 2) than foams produced through fireground foam-making equipment.

The test equipment only produces primary aspirated foam for use during the standard fire tests referred to in this Chapter of the Manual; there are currently no standards available for directly determining the suitability of foam concentrates for fire service secondary aspirated use.

3.3.8 Under What Conditions are the Fire Tests Performed?

Fuel, foam solution, air and fuel temperatures should all be tightly controlled in order for the fire tests to be repeatable and to enable the results to be satisfactorily compared with previous tests. Large variations in temperature can lead to very different extinction and burnback results. Cooler temperatures are likely to lead to quicker extinction times and longer burnback performances.

Wind speed also needs to be carefully controlled, little or no wind will help to produce better, more reproducible tests and results – indoor tests are preferred.

3.3.9 What Burnback Test is Used?

In order to test the security of the foam blanket, a burnback test is required. Burnback tests, where the burnback flames are near to, or actually impinge on, the foam blanket are much more testing. Burnback tests which also involve a burning fuel in a metal container can help to assess the sealing capabilities of foam blankets against very hot materials.

3.3.10 When are the Fire Tests Carried Out?

Are the fire tests only carried out when the foam concentrate is initially tested for compliance with the requirements of a standard or are they carried out on a regular basis (i.e. each manufactured batch/quality control)? Are/were the fire tests carried out by an independent test house or were they carried out by the manufacturer?

Regular fire testing can indicate the continuing suitability of foam concentrates for that task. Some standards only require the fire tests to be carried out once, at the approval stage. Conformance with the standards is then only checked via physical property tests – probably by the manufacturer.

3.4 Periodic Testing of Foam Concentrates

3.4.1 General

Storing foam concentrates as recommended by the manufacturers and as described in Chapter 2, Section 2.2, will help to maintain them in a usable condition. However, no matter how well they are stored, deterioration will take place. Consequently, it is important that samples of stored foam concentrates are tested periodically (e.g. annually) to ensure that they have not significantly deteriorated and that they remain able to effectively extinguish fires.

There are a number of ways of having periodic testing carried out, these include:

- Carry out testing at brigade level.
- Return a sample to the supplier.
- Send a sample to an independent laboratory.

The range of tests that should be carried out to evaluate the condition of foam concentrates requires some specialised equipment and technical expertise. It would not be cost effective or practical for individual brigades to carry out the few tests that would be required each year.

Most foam concentrate manufacturers will carry out this type of testing for a fee. However, some organisations consider it undesirable to rely on manufacturers tests when the manufacturer has a clear commercial interest in the outcome. Whilst there is no suggestion that any supplier has falsified results, it is always possible that an individual could act upon misplaced zeal in the future.

Some manufacturers will test any foam concentrate, not just those they produce. Consequently, if funds allow, it may be advisable to send samples to more than one manufacturer in order to obtain several test reports for comparison.

Independent test houses offer an alternative means of having foam concentrates tested. However, before allowing them to carry out work, always ensure that they have previously analysed foam concentrates and that they can carry out the full range of tests to the required standard.

In order for the amount of deterioration that has taken place to be quantified, it is necessary to have:

- The manufacturer's data sheet (from the time of purchase) for the particular foam concentrate to be tested.

- The results of the routine quality control tests originally carried out by the manufacturer during production on the particular batch or batches of foam concentrate to be tested. This information will normally include the results of physical property tests and, in some instances, the results of fire tests. All manufacturers gather quality control test data during production and they will normally make it freely available on request at the time of purchase. **However, in order to make the best use of this information, it is extremely important that batch numbers are recorded on storage containers and accurate records of usage are kept**.

If the foam concentrate complies with a particular standard, then the limits specified within the standard can also be used to determine whether the foam concentrate still complies with the standard.

As long as the same test methods and equipment are used, the results of periodic testing of stored foam concentrates can be compared with the limits set out in the manufacturers data sheets and with the actual performance of the foam concentrate when originally produced. Any discrepancies can then be identified and investigated further.

It should be remembered that foam concentrates are only part of the equipment and resources necessary to produce effective firefighting foams. Consequently, the whole foam-making system, including the induction or injection equipment, pumps, typical hose lengths, procedures etc. should all be periodically checked individually, and as a whole system, to ensure that all are operating correctly and ultimately providing finished foam of the required quality.

3.4.2 Collection of Foam Concentrate Samples

Foam samples sent for analysis must be representative of the contents of the container from which they have been taken. Samples can be taken as follows:

- *One sample*
  From the bottom of the container only, or from anywhere in the container after thoroughly mixing the contents.

- *Two samples*
  One from the top of the container and one from the bottom.

- *Three samples*
  One from the top, one from the middle and one from the bottom of the container.

Samples should be collected in clean, seal-able containers. Each sample should be at least 1 litre and should completely fill the container. Once the samples have been collected, the collection containers should be sealed and labelled with the date and details of where the sample was taken from. At least two samples should be taken from each sampling location. One sample should be sent to the testing organisation and the other should be kept for further testing should this be required.

Do not write on the sample container the type and concentration of foam concentrate that is in the container, the testing organisation should be able to determine this from the results of their tests. If this information differs from the actual contents then it is an indication that further investigations or tests may need to be carried out to identify the cause of the discrepancy.

Care should be taken when collecting from the bottom of a container due to the possible accumulation of sediment from rust and degradation products. This sediment should be prevented from entering the sample container as it may lead to test results that are not representative of the whole contents of the container.

If only one sample is to be tested, then it is preferred that this should be drawn from the container after the contents have been thoroughly mixed together.

3.4.3 Typical Physical Property Tests

Although a wide range of tests may be carried out, typically, the following physical property tests will be included when manufacturers and test houses determine the condition of stored foam concentrates:

(a) Specific Gravity (Relative Density)

The limits of specific gravity for foam concentrates are normally stated in the manufacturers data sheets. Specific gravity measurements that are higher than the manufacturers limits indicate that the foam concentrate has become more concentrated, probably due to evaporation. Measurements that are below the manufacturers limits indicate that the foam concentrate may have been diluted by water in storage, dilutions of greater than 10% may require that all of the foam concentrate in the container be replaced. Changes in the specific gravity of foam concentrates may also indicate dilution or contamination by other substances. See this Chapter, Section 3.2.2 for more information on specific gravity.

(b) pH (Acidity/Alkalinity)

The limits of pH for foam concentrates are normally stated in the manufacturers data sheets. pH values outside of these limits can indicate that the foam concentrate has been contaminated in some way (e.g. mixed with other foam concentrates), has been broken down by micro-organisms and/or has degraded due to incorrect storage. See this Chapter, Section 3.2.3 for more information on pH.

(c) Sediment (Sludge)

Sediment will tend to sink to the bottom of containers when stored over a long period of time. Consequently, care should be taken when obtaining samples from the bottom of a container to ensure that a representative sample is obtained (see above). The maximum sediment content of a foam concentrate is normally stated in the manufacturers data sheet.

When stored correctly, foam concentrates should only contain very small amounts of sediment. High levels of sediment can indicate that the foam concentrate has been contaminated in some way (e.g. mixed with other foam concentrates), has been broken down by micro-organisms and/or has degraded due to incorrect storage. See this Chapter, Section 3.2.4 for more information on sediment.

(d) Spreading Coefficient

Film-forming foams which no longer provide a positive spreading coefficient when measured have either been contaminated or have significantly degraded. See this Chapter, Section 3.2.5 for more information on spreading coefficient.

3.4.4 Periodic Fire Tests

Any fire tests of stored foam concentrate samples that are carried out by manufacturers or independent test houses are likely to involve significant cost. However, it should be remembered that the main reason for using foam concentrates is to extinguish fires and so this type of testing is the best way of determining whether the foam concentrate remains suitable for its purpose.

Although the physical property tests discussed above will indicate possible changes in the consistency of the foam concentrates, it is the firefighting performance that is of most interest. If the physical property tests indicate a problem, then a fire test should be considered in order to investigate the effects of this on the firefighting performance of the foam concentrate.

The fire tests performed by manufacturers and test houses on a routine basis are generally based on, or around, methods and equipment specified in foam concentrate standards. Typical of this is the $0.25m^2$ area tray fire test specified within the UK Ministry of Defence (MoD) foam concentrate standards (see this Chapter, Section 3.1). This size of fire test is also recommended for quality control use during foam concentrate production in the British, European and International standards for firefighting foam concentrates (see this Chapter, Section 3.1). However, the main difference is that the MoD tests involves the use of avgas or avtur as fuel and the British, European and International standards use heptane (see this Chapter, Section 3.3 for information on test fuels). In order to make the best use of fire test information it is necessary to have previous fire test data available so that true comparisons can be made. For instance, if batch fire test data was available for the foam concentrate when originally purchased then, as long as the same fuel, test methods and equipment are used when testing the stored foam concentrate, the fire test results can be compared for obvious differences in performance. If original fire test data is not available, but the foam concentrate conformed to a particular foam standard when produced, then that standard fire test could be carried out to determine whether the stored foam concentrate still complies with that standard.

Firefighting Foam – Technical

Chapter 4 – The Properties of Finished Foams and The Effects of These on Firefighting Performance

4.1 General

In Chapter 2, the various types and properties of foam concentrates were discussed. This Chapter explains some of the more important properties of finished foams. These properties can greatly affect the firefighting performance of finished foams in terms of:

- **Flame knockdown:** the ability of the finished foam to quickly knockdown flames and control the fire.

- **Extinction:** The ability of the finished foam to extinguish the fire.

- **Burnback resistance:** the ability of the finished foam, once formed on the fuel, to stay intact when subjected to heat and/or flame.

The properties discussed in this Chapter include:

- **Working:** the effort required in mixing air with the foam solution to produce a usable finished foam.

- **Foam flow/fluidity:** the ability of the finished foam to flow over the surface of a fuel and around obstructions.

- **Film formation:** the ability of the finished foam to form a film that spreads over **some** hydrocarbon liquid fuels.

- **Fuel tolerance:** the ability of the finished foam to resist mixing with, and hence contamination by, the fuel.

- **Edge sealing:** the ability of the finished foam to seal against hot metal surfaces.

- **Foam blanket stability/drainage time:** an indication of how well the finished foam blanket retains its liquid content and hence how 'stable' and long lasting it is.

Also included are:

- the suitability of finished foams for base injection,

- finished foam quality,

- the compatibility of various finished foams with each other and with dry powders,

- the typical firefighting characteristics of each of the individual types of foam identified in Chapter 2, particularly when used on liquid hydrocarbon fuel fires.

It should be remembered that other factors, such as type of fuel, equipment and application methods, also have a considerable effect on the performance of finished foams. These areas are discussed in later Chapters and the operational aspects of applying foam are discussed in Volume 2 of the Manual.

4.2 Working

"Working" refers to the action of the internal parts of foam-making equipment on the foam solution stream as it passes through the equipment. The internal parts can include gauzes and baffles which obstruct the flow of the foam solution and greatly assist in the mixing in of air. This helps to produce uniform sized, stable, foam bubbles of acceptable drainage and expansion characteristics.

Some manufacturers claim that, for some low expansion foams such as P and FP, complete formation of stable foam bubbles should take approximately 1/30th of a second. However, foam solution does not begin to form bubbles until it hits the side walls or obstructions approximately half way along the length of the branch. Consequently, it is claimed, the foam solution should be in the branch for a total of 1/15th of a second to form stable finished foam. For main line use at flows of approximately 225 lpm, a low expansion foam-making branch in excess of 1 metre in length would be required to give the required pass through time.

Some foam solutions produce bubbles more readily than others. For instance, SYNDET, AFFF and FFFP foam solutions require less working and hence foam of adequate quality can be produced using shorter branches than are required by P or FP foam solutions. Ultimately, if foam working is excessive, the foam becomes very stiff and loses its flow qualities; for film-forming foams, this may impair there ability to produce an aqueous film on the surface of hydrocarbon liquids. If not enough working is achieved, the foam will be very quick draining, have poor stability and be made up of foam bubbles of irregular size.

Working slows down the foam stream within the branch due to the energy required to produce foam. Consequently, the more a foam is worked within a branch, the less the distance it can be projected.

4.3 Foam Flow/Fluidity

Finished foams that rapidly flow across the surface of fuels and around obstructions can lead to quick flame knockdown and control of a fire. This can be particularly important in aircraft or vehicle crash fire situations where there is a significant risk to life.

Critical shear strength is a measure of the degree of 'stiffness' of finished foam and gives an indication of its ability to flow. Shear strength is measured by a paddle type torsion wire viscometer. These are specialist items of equipment and are not suitable for routine fire service use. Shear strength figures can only be reliably compared if the same type of measuring equipment and measurement methods are used. However, these measurements do not provide a reliable indication of the firefighting capability of foams.

Protein and fluoroprotein foams tend to be stiffer and hence they give higher shear strength measurements than SYNDET, AFFF, AFFF-AR, FFFP and FFFP-AR finished foams. However, the shear strength of finished foam also depends on the amount of working provided by the branch used to produce the foam (see above). Secondary aspirated equipment will produce foam of low shear strength while primary aspirated equipment will produce foam of significantly higher shear strength. In addition, in primary aspirated equipment, the more working that takes place, the higher the shear strength of the finished foam

4.4 Film Formation

The term film formation is often used and applies to AFFF, AFFF-AR, FFFP and FFFP-AR foam concentrates. Under certain conditions, the foam solutions and finished foams produced from these foam concentrates have the ability to produce an aqueous film which spreads over the surface of **some** liquid hydrocarbon fuels. On these particular fuels, the film is said to help cool the surface of the burning liquid to reduce the hydrocarbon evaporation rate, seal in the vapour at the surface of the fuel and hence deplete the supply of fuel to the flames. Consequently, they may assist in the knockdown and extinction of fires in these particular fuels.

The fluorocarbon surface active agents and foaming agents that combine to produce film-forming foams produce a foam solution that has a very low surface tension. This allows a thin film to be formed on, and to spread across, some liquid hydrocarbon fuels. The main factor which influences the effective formation of this film on a hydrocarbon is the surface tension of that hydrocarbon. Film-forming foams tend to be much more effective on liquid hydrocarbons that have a much higher surface tension than the foam solution. High surface tension fuels include kerosene, diesel oils and jet fuels.

The aqueous films produced are extremely thin, typically less than a quarter of a millimetre thick, and are unlikely to form on the surfaces of any hot fuels. Some research carried out in America has indicated that film formation does not occur on aviation gasoline when at temperatures above

60°C. Consequently, these thin films are unlikely to help in extinguishing fires in many flammable fuels that have had long preburns.

It must be stressed that film formation does not take place on all hydrocarbon fuels. In such cases, these foams must rely on the normal extinguishing mechanisms of foam blankets. That is to exclude air from the fire, reduce evaporation and generally cool the fire. This may require more foam to be applied, for a longer period of time than would normally be expected when using a film-forming foam.

It is important to note that although alcohol resistant foams produce aqueous films on **some** liquid hydrocarbon liquids, they do not produce them on water-miscible liquids.

As mentioned above, the ability of a foam to form a film on a hydrocarbon liquid can be determined by measurements of the surface tensions of the foam solution and the hydrocarbon liquid. These measurements are usually carried out in a laboratory. However, in firefighting situations, the conditions are likely to be very different. This makes the conclusions of laboratory measurements generally inapplicable to most practical applications of film-forming foams (see Chapter 3, Section 3.2.5).

Film formation is a very controversial area of firefighting foams. Some firefighters insist that fires can be seen to be controlled and extinguished well ahead of any foam blanket formed; others say that they have seen no evidence of the effects of film formation.

Aqueous films offer little or no burnback protection and, in any case, it can be impossible for firefighters to see where the transparent surface film remains intact and where it has been broken.

The manufacturers of film-forming foam concentrates often state that they may be used primary aspirated, secondary aspirated or non-aspirated for application against hydrocarbon liquid fuel fires.

Petrol fire tests carried out using UK fire service equipment and tactics (Reference 3) found that primary aspirated film-forming foams extinguished the fires in half of the time taken by the same foams used secondary aspirated. Also, the burnback performances of the primary aspirated foams were vastly superior to those of the secondary aspirated foams.

The thinness of the film, and the uncertainty of its formation, makes film-forming foams unsuitable for vapour suppression unless a thick foam blanket is also present. For vapour suppression, primary aspirating equipment will provide a better protective foam blanket than secondary or non-aspirating equipment.

Some foam manufacturers say adequate vapour suppression can be achieved using secondary aspirating equipment with film-forming alcohol-resistant foam concentrates. However, they claim that these should be used at 2 to 3 times their recommended concentration for application to hydrocarbon liquids (e.g. used at 9% concentration instead of their recommended 3%). However, most brigades are unlikely to have equipment capable of proportioning at rates higher than 6%.

It should be noted that the standard film-forming foam concentrates (i.e. AFFF and FFFP) form foam blankets that drain rapidly in order to quickly form films on the fuel surface. Consequently, these foam blankets will need to be replenished at very frequent intervals if adequate vapour suppression is to be maintained. Primary aspirated alcohol resistant film-forming foams require less frequent replenishment due to their much longer drainage times.

4.5 Fuel Tolerance

Fuel tolerance describes how resistant a foam is to mixing with a fuel during application. In general, foams should be applied as gently as possible to the surface of a fuel to reduce the amount of mixing that takes place. Plunging a foam stream directly into a fuel will cause fuel to be mixed in with the foam. If a fire is present, then it is inevitable that this foam and fuel mixture will burn causing partial destruction of the foam blanket. However, some foams are more resistant to mixing with fuel than others.

P foams have poor fuel tolerance and hence suffer from severe fuel contamination when vigorously applied to a fuel. This is because the surface tension properties of protein foam allows fuel to spread over and within the blanket. This can result in burning within the blanket continuing over a long period of time.

The fuel tolerances of FP and FFFP foams are considerably better than that of P foams. This is due to the addition of fluorocarbon surface active agents, which are oleophobic (i.e. they repel oil) and have a very low surface tension. These properties help to resist the spread of fuel across foam bubbles and hence increases their fuel tolerance.

In the case of synthetic detergent based foams, the hydrocarbon surface active agents that are used in their formulation tend to emulsify oils with water. This causes the foam to pick up large quantities of fuel which can readily ignite. Fuel tolerance has been improved in the case of AFFFs and AFFF-ARs by the additional use of a high proportion of fluorocarbon surface active agents.

In contrast to the above, fuel emulsifiers (see Chapter 2, Section 2.1.8) are oleophilic (i.e. they attract oil) and rely on mixing well with fuel in order to form an emulsion. The emulsion is claimed by the manufacturers to consists of fuel molecules encapsulated in water molecules. This, they say, significantly reduces the amount of vapour released by the fuel making the mixture incapable of sustaining combustion. The vigorous application of emulsifiers directly to fires in petroleum based fuels is claimed by the manufacturers to result in rapid control and extinction. In addition, because the fuel molecules have been encapsulated, they say that it is unlikely that re-ignition will occur. Emulsifiers have only recently been introduced and their performance relative to other foam concentrates and firefighting media has yet to be proven in the UK.

4.6 Edge Sealing

The term edge sealing relates to the ability of a foam blanket to seal against hot metal surfaces. Hot metal surfaces can cause breakdown of a foam blanket due to the boiling off of its water content and increased vapour release from the fuel at the hot surface. This can result in the inability of a finished foam to fully extinguish fires at this interface.

Foams which have a good resistance to heat tend to exhibit good extinguishing performances and burn back resistance and therefore should have good edge sealing properties. However, when hot metal surfaces (i.e. in excess of 100°C) are encountered by a foam blanket, destruction of the foam blanket is inevitable and steps should be taken where possible to cool these surfaces sufficiently to ensure edge sealing can take place. This can be particularly important when fighting large tank fires.

4.7 Foam Blanket Stability/ Drainage Time

Drainage time is a measurement of the rate at which foam solution drains out of finished foam and hence provides an indication of the stability of the foam blanket. Drainage time is often used to compare the quality of various finished foams, however, it does not provide a reliable indication of the firefighting capability of foams.

A long drainage time, and hence slow loss of water from the finished foam, tends to indicate that the finished foam is capable of maintaining its stability and heat resistance. This is usually the case with most P, FP, AFFF-AR and FFFP-AR foams. However, this is not true for low expansion SYN-DET foams which generally produce finished foams with long drainage times but have very poor heat resistance.

A short drainage time tends to indicate that the finished foam loses its water content quickly and renders it vulnerable to high temperature flame and hot surfaces. AFFFs and FFFPs tend to have low drainage times and poor heat resistance.

The drainage times of finished foams depends not only on the foam concentrate but also on the foam-making equipment used to produce it. Secondary aspirated equipment will produce finished foams with short drainage times while primary aspirated equipment will generally produce finished foams with significantly longer times. In addition, in primary aspirated equipment, the more working that takes place, the longer the drainage times.

Drainage for low expansion foams is usually expressed as 25% drainage time. This is defined as the time taken for 25% of the original foam solution content (by volume) to drain from the finished foam. For medium and high expansion foams, 50% drainage times are normally given.

Figure 4.1 shows the basic principles of measuring low expansion foam expansion ratios and drainage times. The current British Standards for foam concentrates (see Chapter 3) should be referred to for exact details of equipment and test methods to be used. Expansion ratios and drainage times of finished foams can only be reliably compared if the same type of foam concentrate, measuring equipment, foam-making equipment and measurement methods are used. In particular, the height of the measurement container has a significant impact on the length of drainage time measurements; short containers give short drainage times, tall containers give longer drainage times.

Firefighters should remember that when a foam drains, its volume will seem almost unchanged. Although its integrity may appear good, its fire resistance will be low as it will have lost much of its foam solution content.

4.8 Vapour Suppression

It is extremely important that foam blankets prevent fuel vapour percolating through to their upper surface. If the foam blanket is unable to prevent this, then it is likely that the vapour will continue to burn on the surface of the foam. This can quickly lead to the complete destruction of the foam blanket.

4.9 Burnback Resistance

Burnback resistance is the ability of a foam blanket to resist destruction from direct contact with heat and flames. Such contact occurs during initial foam application where the foam blanket will be continually covering, and moving against, flame. It can also occur, once successful foam application has been achieved, from a small area of sustained burning or from a new ignition source.

Burnback resistance is one of the main properties assessed when testing the firefighting performance of foams. Usually, once a test fire has been extinguished, the burnback resistance of the foam

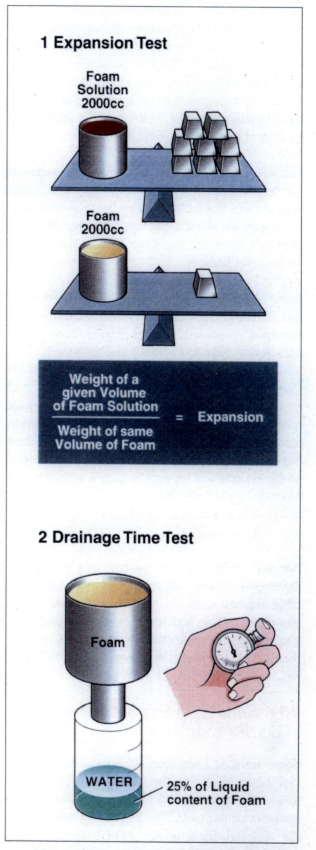

Figure 4.1 Diagram showing tests to determine foam properties.
  (1) Expansion test.
  (2) Drainage time test.

blanket is assessed. Either a small area of foam is removed and the fuel underneath is re-ignited or, a flame is continuously played on to a small area of the foam blanket. The measurement made is known as the burnback time. This is the time taken from re-igniting the fuel, or applying flame to the foam blanket, until the re-involvement in flame of an area of the surface of the fuel.

Often, it is the 25% burnback time that is quoted for the burnback resistance of foams. This is the time it takes for a 25% area of the fuel surface to become re-involved in flames. The longer the 25% burnback time, the better the burnback resistance of the foam blanket.

Some foams, such as P, FP and the alcohol resistant film-forming foams have significantly greater burnback resistance, and hence give longer burnback times, than AFFF, FFFP and SYNDET.

Generally, the more foam applied to a fire after extinction has occurred, the better the burnback resistance will be. However, if a foam blanket is left over a period of time and allowed to drain without being replenished, the burnback resistance of the blanket will be significantly impaired.

4.10 Water-miscible Fuel Compatibility

Alcohol resistant foam concentrates have been developed to deal with fires involving water-miscible liquids such as alcohols and some petrol blends containing high levels of alcohols and other similar fuel performance improvers. These, and the finished foams that they produce, are described in Chapter 2, Section 2.1.4.

4.11 Suitability For Subsurface (Base) Injection

Some finished foams can be introduced, via special equipment, into the bases of large storage tanks. The foam then floats to the surface of the contents of the tank. This has the advantage that the finished foam is not carried away by the updraught created by large fires and is not deteriorated by flames on the way to the surface of the fuels. However, foams that are used for subsurface injection need to have a high tolerance to fuel contamination otherwise the foams would burn away immediately on contact with the flames on the surface of the product.

Subsurface injection can only be used in tanks containing certain hydrocarbon fuels; it cannot be used for tanks containing water-miscible fuels because, even with alcohol resistant foams, these fuels will destroy the foam blanket on contact and a foam blanket will not form. In addition, this will mean that the polymeric skin cannot form on the surface of the fuel (see Chapter 2, Section 2.1.4).

FP, FFFP, FFFP-AR, AFFF and AFFF-AR foams are generally considered suitable for base injection.

4.12 Quality of Finished Foam

The production of good quality finished foam depends on:

- the use of a suitable type and quality of foam concentrate for the task in hand;
- foam concentrate in good condition due to correct storage;
- foam concentrate used at the correct concentration;
- good design and choice of equipment;
- good maintenance of equipment;
- correct pump pressure and foam solution flow for the equipment in use.

Drainage times and expansion ratios (and sometimes shear strength) can be measured and compared to provide an immediate indication of the 'quality' of a finished foam. Often, firefighters will look at and feel the finished foam produced by their equipment and give an immediate assessment of its quality. 'Wet' foams, i.e. those with short drainage times, are often referred to as being of poor quality while those that are 'dry', i.e. those with long drainage times, are referred to as being of good quality.

However, there is no overall definition of a 'good' quality foam. This really depends on which foam is being assessed and its intended use.

On some fuels, AFFFs and FFFPs rely on being 'wet' to assist in the formation of a film on the surface and for quick cooling. It can also assist in giving them quick control and knockdown capabilities. P and FP finished foams are often better used 'dry' to provide acceptable knockdown and extinction performance and good burnback resistance. If used too 'wet', by applying them via poor foam-making equipment for instance, these foams are likely to give very poor firefighting performance. Applying them too 'dry' will result in very thick foam being produced which does not flow very easily and again results in very poor firefighting performance.

As part of the routine checking of the operation of a foam-making system, expansion ratios and drainage times of the finished foam can be measured and compared with previous measurements.

Figure 4.1 shows the basic principles of measuring low expansion foam expansion ratios and drainage times. The current British Standards for foam concentrates (see Chapter 3) should be referred to for exact details of equipment and test methods to be used. Drainage times and expansion ratios can only be reliably compared if the same type of foam concentrate, measuring equipment, foam-making equipment and measurement methods are used.

4.13 Compatibility of Finished Foams

4.13.1 With Other Finished Foams

Generally speaking, all types of finished foam can be used together on a single fire, although the order of application may affect their performance. For example, film-forming foam would be better applied first for a quick knockdown and extinction of a hydrocarbon fuel fire followed by an application of FP foam to provide good burnback resistance. Applying these foams in reverse order would result in the partial breakdown of the FP foam blanket, and hence reduced burnback resistance, due to the film-forming foam blanket quickly draining with the resulting falling liquid droplets bursting the FP foam bubbles.

4.13.2 With Dry Powder

Some finished foams will react unfavourably with certain fire extinguishing powders if used at the same incident. The manufacturer should be asked whether there are any particular incompatibles to their product. Firefighters should remember to consult the industrial/MOD/CAA brigades etc., in their areas, as well as neighbouring local authority brigades where appropriate, to find out what dry powder types they are using. Foam concentrate manufacturers should then be contacted for advice on compatibility.

4.14 Typical Characteristics of Finished Foams

4.14.1 General

The following Sections highlight the typical characteristics of low expansion finished foams produced from each of the main types of foam concentrate described in Chapter 2. These characteristics relate mainly to their use on hydrocarbon liquid fuel fires although other comments are made concerning, for instance, their compatibility with water-miscible fuels. The terms used here have been explained earlier in this or the previous Chapters, see also the Glossary of Terms.

Table 4.1 overleaf enables a quick comparison to be made of the typical firefighting related characteristics of low expansion finished foams made from each of the main foam types. The contents of this table are intended to provide information on typical performance during general fire service use, in particular, when used against hydrocarbon spill fires.

The table should be read in conjunction with the contents of the remainder of this Section which provide more details of the characteristics for each foam type. In addition, some comments regarding the suitability of different foam concentrate types for use in tackling storage tank fires are given in Volume 2 of the Manual.

It should be remembered that there are many companies manufacturing each of the different foam concentrate types. The quality of foam concentrates produced will vary from manufacturer to manufacturer and often different quality versions of the same foam type will be available from the same manufacturer. Consequently, the following Sections indicate the typical characteristics of finished foams produced from each of the foam types.

Table 4.1: *Typical Characteristics of Low Expansion Finished Foam*

CHARACTERISTIC	FOAM TYPE						
	P	FP	FFFP	FFFP-AR	SYNDET	AFFF	AFFF-AR
Requires to be well 'worked'?	Yes	Yes	No	No	No	No	No
Foam Flow/Fluidity	▫	▫▫▫	▫▫▫▫	▫▫▫▫	▫▫▫▫	▫▫▫▫	▫▫▫▫
Film-forming on **some** hydrocarbon liquids?	No	No	Yes	Yes	No	Yes	Yes
Hydrocarbon Fuel Tolerance	▫	▫▫▫▫	▫▫▫	▫▫▫▫	▫	▫▫▫	▫▫▫▫
Flame Knockdown	▫	▫▫▫	▫▫▫▫	▫▫▫▫▫	▫▫▫▫▫	▫▫▫▫▫	▫▫▫▫▫
Edge Sealing	▫▫▫	▫▫▫▫	▫▫	▫▫▫▫	▫	▫▫	▫▫▫▫
Extinction	▫	▫▫▫	▫▫▫▫▫	▫▫▫▫	▫▫	▫▫▫▫▫	▫▫▫▫
Foam Blanket Stability/Drainage Time	▫▫▫▫	▫▫▫	▫▫	▫▫▫▫	▫▫▫▫▫	▫▫	▫▫▫▫
Burnback Resistance	▫▫▫▫	▫▫▫▫▫	▫▫	▫▫▫▫	▫▫	▫▫	▫▫▫▫
Vapour Suppression	▫▫▫▫	▫▫▫▫▫	▫▫	▫▫▫▫	▫▫	▫▫	▫▫▫▫
Foam Application	LX	LX MX	LX MX SA	LX MX SA	LX MX HX	LX MX SA	LX MX SA
Water-miscible Fuel Compatible?	No	No	No	Yes	No	No	Yes
Suitable for Hydrocarbon Subsurface Injection?	No	Yes	Yes	Yes	No	Yes	Yes

Notes to Table 4.1:
This table summarises the typical characteristics that can be expected from good quality low expansion finished firefighting foams when used to fight some flammable hydrocarbon liquid fuel spill fires. The characteristics of finished foam will vary depending on factors such as fuel, application technique, equipment and the quality of the foam concentrate used. The firefighting performance contents of this table are based on the results of work carried out on petrol spill fires (Reference 5). A difference in performance of one grade is not significant due to the tight cut off points in the results used to generate the grades and the level of repeatability of the tests. However, where there is a difference in performance of two or more grades, the difference is significant.

▫▫▫▫▫ = Very Good; ▫▫▫▫ = Good;
▫▫▫ = Acceptable; ▫▫ = Poor;
▫ = Very poor; LX = Low Expansion;
MX = Medium Expansion; HX = High Expansion;
SA = Secondary Aspirated

The firefighting performance contents of this table are based on the results of work carried out by the Home Office FRDG on petrol spill sites.

Good quality foam concentrates may have better characteristics, those of bad quality foam concentrates may be considerably worse. Obviously, other factors such as fuel, application technique and the type of equipment used will also greatly affect these characteristics.

4.14.2 Individual Foam Characteristics

(a) P

Low expansion finished foams produced from P foam concentrates tend to have the following useful characteristics:

- provide acceptable sealing against hot metal surfaces;
- form stable foam blankets with slow foam drainage times;
- good burnback resistance;
- good vapour suppression;

and the following disadvantages:

- can be used to produce low expansion foam only;
- require to be well worked to make acceptable finished foam, they must be used primary aspirated;

- very slow flowing and stiff, protein foams do not quickly reseal breaks in the foam blanket or seal around obstructions. These are some of the major reasons for the slow fire knockdown and extinction performance of protein foams;
- very poor fuel tolerance when applied forcefully to the surface of a fuel. This is the main reason for very slow fire knockdown and extinction performances;
- unsuitable for use with water-miscible fuels;
- unsuitable for subsurface (base) injection.

(b) FP

Low expansion finished foams produced from FP foam concentrates tend to have the following useful characteristics:

- flow quicker than P foams over fuel surfaces, reseal breaks in the foam blanket and seal around obstructions. These properties assist in producing fire knockdown and extinction performances that are quicker than that achieved by P;
- good fuel tolerance so they can be applied reasonably forcefully if absolutely necessary;
- produce acceptable fire knockdown and extinction performance although generally slower than film-forming foams;
- good sealing properties against hot metal surfaces;
- form stable foam blankets with slow foam drainage times;
- very good burnback resistance;
- very good vapour suppression;
- suitable for subsurface (base) injection;

and the following disadvantages:

- do not flow as well as film-forming foams. This often results in slower knockdown and extinction performances when compared to those of film-forming foams;
- require to be well worked to make acceptable finished foam, they must be used primary aspirated;
- unsuitable for use with water-miscible fuels although alcohol resistant FP is available for certain specialised applications.

(c) FFFP

FFFPs were designed to exhibit a combination of AFFF and FP characteristics. The intention was to produce a foam concentrate that had the knockdown and extinction performance of AFFF combined with the good burnback resistance characteristics of fluoroprotein. However, fire tests (Reference 5) have indicated that although low expansion FFFP gives similar firefighting and burnback performance to AFFF, the burnback performance is greatly inferior to that achieved by fluoroprotein and is generally not much better than AFFF.

Low expansion FFFP finished foams tend to have the following useful characteristics:

- usable foam can be produced with minimal working, manufacturers suggest that they can be used primary and secondary aspirated;
- flow quicker than P and FP foams over liquid fuel surfaces, quickly reseal breaks in the foam blanket and flow around obstructions. This often results in very quick fire knockdown and extinction. On **some** liquid hydrocarbon fuels, these characteristics may be enhanced by the film-forming capabilities of FFFP;
- suitable for subsurface (base) injection;
- moderate resistance to fuel contamination although not as fuel tolerant when used on non-water-miscible fuels as alcohol resistant film-forming foams or FP foams;

and the following disadvantages:

- poor at sealing against hot objects;
- poor foam blanket stability and very quick foam drainage times;
- poor burnback resistance;
- poor vapour suppression;
- unsuitable for use with water-miscible fuels.

(d) Synthetic (SYNDET)

SYNDET finished foams are versatile in that they can be used for firefighting at low, medium and high expansion. In the UK, they are mainly used at medium and high expansion foams.

The following comments mainly relate to their use at low expansion in order to enable a comparison to made with all of the other foam types discussed. However, many of these comments are also relevant for their use at medium and high expansion.

Low expansion SYNDET finished foams tend to have the following useful characteristics:

- produce acceptable foam with minimal working, must be used primary aspirated;
- quick-flowing which can assist in producing quick fire knockdown. Medium and high expansion SYNDET foams do not flow as readily, however, the large volume of foam produced can achieve quick knockdown and extinctions;
- very stable foam blankets with very slow foam drainage times. Medium and high expansion SYNDET foams can be severely affected by wind.

they have the following disadvantages:

- very poor resistance to fuel contamination, often resulting in poor extinction and burnback performance. Medium and high expansion applications of SYNDET are relatively gentle and so fuel contamination is less of a problem;
- very poor sealing around hot objects often resulting in poor extinction performances;
- poor burnback resistance;
- poor vapour suppression capabilities at low expansion; vapour suppression characteristics much improved at medium and high expansion;
- unsuitable for use with polar fuels;
- unsuitable for subsurface (base) injection.

(e) **AFFF**

Low expansion AFFF finished foams tend to have the following useful characteristics:

- usable foam can be produced with minimal working, manufacturers suggest that they can be used primary and secondary aspirated;
- flow quicker than P and FP foams over liquid fuel surfaces, quickly reseal breaks in the foam blanket and flow around obstructions. This often results in very quick fire knockdown and extinction. On some liquid hydrocarbon fuels, these characteristics may be enhanced by the film-forming capabilities of AFFF;
- suitable for subsurface (base) injection;
- moderate resistance to fuel contamination although not as fuel tolerant on non-water miscible fuels as alcohol resistant foams or FP foams;

and the following disadvantages:

- poor at sealing against hot objects;
- poor foam blanket stability and very quick foam drainage times;
- poor burnback resistance;
- poor vapour suppression;
- unsuitable for use with polar fuels.

(f) **Alcohol Resistant Foam Concentrates (AFFF-AR and FFFP-AR)**

Low expansion finished foams produced from AFFF-AR and FFFP-AR alcohol resistant foam concentrates tend to have the following useful characteristics:

- suitable for use on fires involving water-miscible liquids such as alcohols and those petrol blends that contain high levels of alcohols and other similar fuel performance improvers;
- suitable for use on hydrocarbon liquid fuel fires;
- usable foam can be produced with minimal working, manufacturers suggest that they can be used primary and secondary aspirated on non-water miscible fuels. On water-miscible fuels, the foam solutions must not be applied non-aspirated and also their use on these fuels when secondary aspirated cannot be recommended;
- flow quicker than P and FP foams over liquid fuel surfaces, quickly reseal breaks in the foam blanket and flow around obstructions. This often results in very quick fire knockdown and extinction. On **some** liquid hydrocarbon fuels, these characteristics may be enhanced by the

film-forming capabilities of AFFF, film-forming does not occur on water-miscible fuels;
- good resistance to contamination from hydrocarbon fuels so can be applied forcefully to these if absolutely necessary. Only gentle application techniques should be used when applying these foams to water-miscible fuels.
- suitable for subsurface (base) injection. They must not be used for base injection into water-miscible fuels;
- When used on non-water miscible fuels, control and extinction times are similar to those of conventional AFFF and FFFP foams with burnback performance similar to that of FP. Extinction and burnback performance is considerably better when used primary aspirated (i.e. using a foam-making branch) than when used secondary aspirated (i.e. using a water branch);
- very stable foam blankets with slow foam drainage times;
- good at sealing against hot metal objects;
- good burnback resistance;
- good vapour suppression.

and the following disadvantages:

- care is required in selecting the correct rate of induction due to the need to use at 3% concentration for hydrocarbon fuels and at 6% for water-miscible fuels. However, some alcohol resistant foams are available that may be used at the same induction rate (normally 3%) for both hydrocarbon and water-miscible fuels.

4.15 Environmental Impact of Firefighting Foams

4.15.1 General

Firefighting foams are the most effective means of extinguishing most liquid fuel fires. In doing so, they greatly reduce fire spread, the air pollution potential of a fire and the amount of water that needs to be used to tackle the fire. This in turn reduces the amount of contaminated water produced during firefighting operations and the environmental impact of this run-off.

Firefighting foams can also be of benefit by preventing the release of flammable or toxic vapour into the environment.

The use of foams for firefighting is infrequent and at changing locations. Consequently, the impact on the environment in these areas does not accumulate although it can be severe at the time of the incident. In contrast, areas used for training are likely to be frequently exposed to contamination by foams and the run-off from these sites should be controlled by containment and disposal to appropriate treatment works.

Generally, the environmental effects of foams are considered in terms of their toxicity and their biodegradability. It should be remembered that it is the total volume of the foam concentrate that is released into the environment that is of concern, it does not matter by how much it has been diluted.

4.15.2 Toxicity

The aquatic toxicity of a substance (i.e. how poisonous it is to water life) is usually measured in terms of its LC_{50}. This is the lethal concentration of the substance in water at which 50% of test specimens die within a fixed time period under test conditions. Generally speaking, the higher the LC_{50} value, the less impact the substance will have on aquatic life.

Sometimes, LC_{10} and even LC_0 measurements are made or required. These are much more demanding with LC_0 indicating the concentration at which there has been no observable affect to the test specimens.

Unfortunately, the range and type of test specimens that are tested varies widely as does their susceptibility to the effects of the substance.

For most foam concentrates, only the foam manufacturers' toxicity information is available; very few independent tests have been carried out. Toxicity testing can be very expensive to perform. Consequently, some foam manufacturers do not provide comprehensive values, others provide values for a small or wide range of test specimens including algae, water flea (often Daphnia Magna) and fish (often either rainbow trout or fathead minnow). However, it is extremely difficult to compare the toxic effects of foam concentrates unless the same specimens, test conditions and toxicity measurement criteria are used.

A review of firefighting foam concentrates carried out by the Water Research Council on behalf of the National Rivers Authority during 1994 (Reference 6) concluded that all of the toxicity data they collected from various sources, particularly manufacturers, indicated that none of the foam concentrates were of high acute toxicity to test specimens. They found that most foam concentrates were tested on water flea or fish although indications were that testing on algae would have produced results for a more sensitive species.

From the data they collected, some SYNDET foam concentrates appeared to be the most toxic and all of the protein based foam concentrates were of low acute toxicity. However, some AFFF and AFFF-AR foam concentrates were also found to be in this low acute toxicity band.

4.15.3 Biodegradability

Biodegradability of a substance is a measure of how quickly it is broken down by bacteria. Bacteria in the environment will break down and eat the substance, extracting oxygen from the surrounding water as they do so.

Measurements of biodegradability are made by carrying out two different tests and comparing their results.

One test provides a measure of the Chemical Oxygen Demand (COD). This is the total amount of oxygen required to degrade a set amount of foam; the lower the COD, the less oxygen that is stripped from the environment.

The second test provides a measure of the Biochemical Oxygen Demand (BOD). This is an indication of the foam concentrate's ability to consume that amount of oxygen within a specified time period, usually 5 days (referred to as BOD_5).

Most of the data issued by foam manufacturers consists of biodegradability values in terms of BOD_5/COD as a percentage. The higher the percentage, the higher the biodegradability of a foam, the quicker the foam is broken down.

The Water Research Council (see above) found that in most environmental hazard assessments, high biodegradability is considered desirable. However, it has been found that the main environmental impact of the use of foam is the rapid depletion of oxygen from water due to high biodegradability. This has the effect of asphyxiating aquatic organisms. They concluded that slower (low) biodegradability of foam concentrates may in fact be more desirable when making future environmental hazard assessments.

They also found that manufacturers only provide limited biodegradability test data which was of little use in differentiating between the biodegradability of different foam concentrates. From the data available, there were indications that foam type was not a good indicator of biodegradation potential. Five different foam types were of low biodegradability, these were SYNDET, P, FP, AFFF and AFFF-AR. However, some AFFF and SYNDET foam concentrates were of high biodegradability.

None of the data gathered enabled an assessment to be made of the biodegradability of the fluorosurfactants contained in AFFF, AFFF-AR, FP, FFFP and FFFP-AR foam concentrates. These chemicals may remain (persist) in the environment for long periods of time before degrading.

Firefighting Foam – Technical

Chapter 5 – Equipment

5.1 General

This Chapter describes some of the foam equipment that is currently in use within the UK fire service. The aim is not to describe every item of equipment available but to give examples and indications of their performance.

The two main types of foam equipment described here are:

- Foam-making equipment (e.g. foam-making branches, foam-making generators etc.);

- Foam concentrate induction and injection equipment (e.g. in-line inductors etc.).

Specialised foam equipment for fighting storage tank fires is not covered here but is described in Volume 2 of the Manual.

Much of the information contained within this Chapter has been obtained from manufacturers. **This information should only be used as a guide to performance and may not reflect actual performance under operational conditions.** Equipment should always be tested under realistic conditions before purchase to ensure that all operational requirements and performance criteria are met. In addition, the induction/injection and foam making equipment should be checked at regular intervals, using operational pressure / flow conditions and hose lengths, to ensure that the foam-making system is working correctly and that the required quality of foam is being produced (see Chapter 4, Section 4.12 and this Chapter, Section 5.5).

5.2 Foam-Making Equipment

5.2.1 General

The primary aspirating foam-making equipment used by brigades can be divided into the following main categories:

- LX hand-held foam-making branches;
- LX hand-held hosereel foam unit;
- LX foam generators;
- LX foam monitors;
- MX hand-held foam-making branches;
- LX and MX hand-held water branch 'snap-on' attachments;
- MX foam pourers;
- HX foam generators.

The above equipment is available in various sizes requiring from less than 50 litres per minute to over 15,000 litres per minute of foam solution.

Some types of foam-making equipment are fitted with a means of picking up foam concentrate at the equipment via a length of tube; these are known as 'self-inducing'. Some types of these operate at fixed induction rates (e.g. 3% or 6%) while others have control valves which enable them to be quickly adjusted to pick-up foam concentrate at a range of concentrations. It is also usually possible to turn off the induction facility completely so that the foam-making equipment can be used with pre-mix foam solutions (see below).

With all other types of foam-making equipment, the foam concentrate must be introduced into the water stream at an earlier stage, usually by some form of induction or injection equipment (see this Chapter, Section 5.3), this results in the production of a 'premix' foam solution. In other words, the foam concentrate and water have been mixed together prior to arriving at the foam-making equipment.

A less often used method of producing a premix foam solution is by mixing the correct proportions of water and foam concentrate in a container prior to pumping. Some brigades have used this method in the water tanks of water tenders.

Secondary aspirated foam is often produced using standard main line and hosereel water branches. However, some purpose designed secondary aspirating LX foam-making branches and monitors have been produced.

Some large output primary and secondary aspirating monitors are described in Volume 2 of the Manual. These are primarily meant for applying foam to storage tanks. The foam solution supply rates for these monitors can be in excess of 40,000 litres per minute.

In general, the means of distinguishing between the capacities of different foam-making equipment is either by the nominal flow requirement of the equipment (litres per minute, lpm) and/or the volume of the foam produced (cubic metres per minute, m³/min). Usually, for LX foam-making equipment, it is the nominal flow requirement only that is used to classify them. The use of this classification also aligns with application rates (see Chapter 7) which recommend the minimum amounts of foam solution, in litres per minute, that should be applied to each square metre of fire area (lpm/m^2).

For MX and HX foam-making equipment, both the nominal flow requirement and the volume foam production are used to classify their output. Generally, the volume foam production figures specified by manufacturers will be those achieved when using SYNDET foam concentrates. However, film-forming foams may also be used to produce MX foam and these are likely to give different foam volume outputs.

5.2.2 LX Hand-held Foam-making Branches

(a) How They Work

Figure 5.3 illustrates the principal features of a typical hand-held LX foam-making branch. Designs vary and will incorporate some or all of these features. The strainer is frequently omitted, as often is the on/off control.

Figure 5.1 FB5x MkII 225 litres/min foam solution at 5.5 BAR. (Photo: Mid and West Wales Fire Service)

Figure 5.2 F450 450 litres/min of foam solution at 7 BAR. (Photo: Mid and West Wales Fire Service)

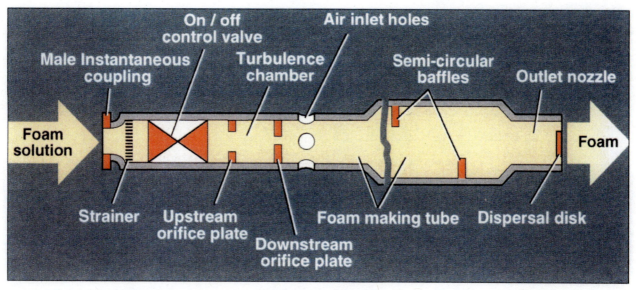

Figure 5.3 Principal features of a Low Expansion foam branch pipe.

In the diagram are two orifice plates. The upstream orifice is the larger of the two and its function is to create turbulence in the space between the two orifice plates so that when the jet issues from the downstream orifice, it rapidly breaks up into a dense spray. The spray fills the narrow inlet section of the foam-making tube and entrains large quantities of air through the air inlet holes. The downstream orifice is smaller and is calibrated to give the designed foam solution flow rate at the recommended operating pressure (e.g. 225 lpm at 7 bar branch pressure).

Most foam-making branches have a narrow section at the inlet end in which the air entrainment takes place, and then a wider section in which the foam forms. The wider section of the foam-making tube sometimes contains 'improvers' (e.g. semi-circular baffles, gauze cones) which are designed to work the foam solution in order to produce longer draining finished foam. The drawback of using improvers is that the extra working of the foam that they cause uses energy from the foam stream resulting in a reduction in the distance that the finished foam can be thrown.

At the outlet, the branch is reduced in diameter to increase the exit velocity, thus helping the finished foam to be thrown an effective distance. The design here is crucial; too narrow an outlet produces back pressure which results in less air entrainment and finished foam of very low expansion ratio and very short drainage times. If the outlet is too large, the expansion is higher but the throw is reduced.

Some branches may also contain flow straightening sections at the nozzle to reduce turbulence at the outlet of the branch. These assist in forming a coherent 'rope' of finished foam with little fall out of foam along its trajectory. However, these tend to considerably reduce the throw of the branch. For foam-making branches without flow straightening sections, considerable amounts of foam can fall out of the stream along their trajectory resulting in a greatly reduced foam volume actually arriving in the area of impact.

(b) **LX Foam-making Branch Performance**

It is generally recognised that the longer the foam-making tube, the better the working and mixing of foam solution with air. This results in a more stable finished foam with drainage times that are longer than those produced by shorter foam-making branches.

Large scale petrol fire trials have been carried out (Reference 7) where the firefighting performance of a short LX foam-making branch (approximate length 0.5m) and a longer LX foam-making branch (approximate length 0.8m) were compared. When these fires were fought with film-forming foam concentrates, both types of branches

Firefighting Foam – Technical 41

Table 5.1: *LX foam-making branches: comparison of foam properties of long and short LX foam-making branches*

Foam-making branch	Foam concentrate	Flow (lpm)	Expansion	25% drainage time
Short	AFFF	225	17.1	1min 43secs
Long	AFFF	225	15.7	2min 35secs
Short	AFFF-AR	225	14.2	3min 15secs
Long	AFFF-AR	225	13.4	6min 4 secs
Short	FFFP	225	11.9	54secs
Long	FFFP	225	13.7	2min 15sec
Short	FP	225	9.8	1min
Long	FP	225	11.6	3min 15sec

Notes to Table 5.1: Measurements taken form References 5 and 7.

produced foams that gave similar knockdown and extinction times but the foam produced by the longer foam branches had much longer drainage times and gave significantly better burnback protection.

During these same fire tests, it was found that the firefighting performance of FP foam was extremely poor when used through the short foam-making branch but perfectly adequate when used through the longer foam-making branch. Details of some of the measurements made of the foam produced by these branches are given in Table 5.1.

LX foam-making branches operating at their recommended pressure (usually either 5.5 or 7 bar branch pressure) with a flow of 225 lpm are claimed by the manufacturers to give throw distances varying from 12 metres (coherent rope) to 21 metres (no internal baffles etc.). Hand-held LX foam-making branchpipes are also available with nominal flow requirements of approximately 450 lpm and 900 lpm at 7 bar branch pressure. These are claimed to throw finished foam a few metres further than the 225 lpm branches.

Some foam-making branches are specially designed for use with film-forming foam concentrates in crash fire situations. These branches have adjustable jaws at the outlet giving the option of a cohesive jet or a fan like spray. They also have an on/off trigger mechanism controlling the release of the foam.

One adjustable jaw type 225 lpm foam-making branch is claimed by the manufacturer to give throws ranging from 7 metres with the jaws closed (i.e. spray mode) to 13 metres with the jaws open (i.e. jet mode) when operated at 7 bar.

5.2.3 LX Hand-held Hosereel Foam Unit

This consists of a portable hand-held unit, similar to an extinguisher (see Figure 5.4), which can contain up to 11 litres of foam concentrate. An appliance hosereel is connected to an adaptor at the top of the unit and water is supplied at between 2 and 10.5 bar.

A small proportion of the water is diverted to fill a completely deflated flexible bag within the container. Inflation of the bag displaces the foam concentrate via a siphon tube, the concentrate entering the main water stream and passing to an integral LX foam-making branch to give a jet of primary aspirated foam. The unit is controlled via an on/off valve on the adaptor.

Figure 5.4 Hand-held Hose reel Foam Unit.

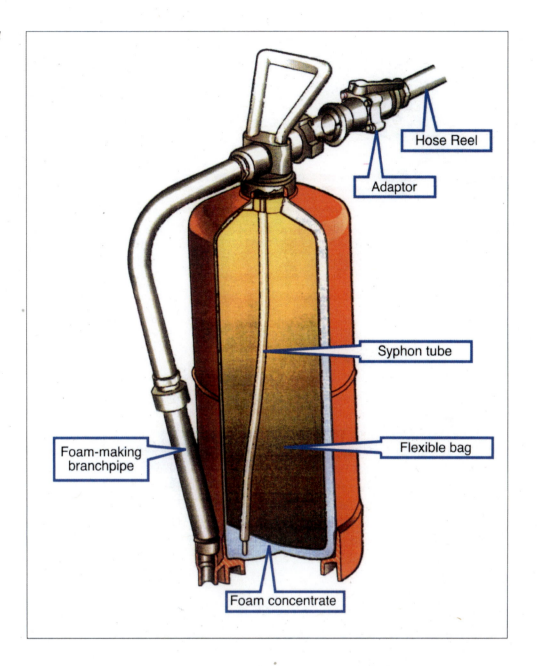

When operated at 3.5 bar with a flow rate of 46 lpm, the manufacturer claims that the unit will produce foam with an expansion of approximately 8.

5.2.4 LX Foam Generators

As an alternative to a foam-making branch, a LX foam generator may be used. This, when inserted in to a line of hose, induces appropriate amounts of foam concentrate and air into the water stream to generate finished foam, which is then delivered through the hose to a water-type branch for application as aspirated foam. The foam concentrate is induced using the same principle as that of an in-line inductor (see below), and the air is drawn in through orifices adjacent to the water inlet. The equipment can only work against limited back pressure, so the length and size of the hose between the generator and branch, and the size of the branch, need to be carefully selected.

Such generators are used to a limited extent in the Fire Service. A typical example has a recommended water inlet pressure of 10.5 bar and a nominal water requirement is 255 lpm. It can be used with up to 60 m of 70 mm hose and a water branch with a 38 mm nozzle. Larger sizes of generator are made but are generally used in fixed installations.

Firefighting Foam – Technical 43

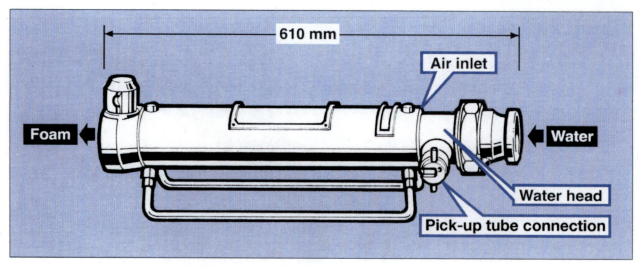

Figure 5.5 Model 5A low expansion foam generator.

5.2.5 LX Foam Monitors

Primary aspirating LX foam monitors are larger versions of foam-making branches which cannot be hand-held. They may be free-standing and portable, mounted on trailers or mounted on appliances. They usually have multiple water connections, and may be self-inducing or used in conjunction with one of the induction methods described in Section 5.3 below. They can also be found in fixed installations at oil-tanker jetties and refineries or as oscillating monitors in aircraft hangars. Similar monitors are fitted to airport foam tenders, often with adjustable jaws which allow the option of a flat fan-shaped spray.

There are numerous LX foam monitors in use coming in a wide range of nominal flow and inlet pressure requirements. One example has a nominal flow requirement of approximately 1800 lpm at an inlet pressure of 7 bar and is claimed by the manufacturer to have a maximum horizontal range of 50 metres and a maximum height of throw of 18 metres. Another example operates at approximate-

Figure 5.6 Portable foam monitor in use.
(Photo: West Midlands Fire Brigade)

44 *Fire Service Manual*

Figure 5.7 Photograph showing the layout of a typical trailer-mounted foam monitor.
(Photo: Angus Fire Armour Ltd.)

ly 4300 lpm at 10 bar inlet pressure with a claimed throw of 60 metres and height of 24 metres.

The throw distances and heights provided by manufacturers are often recorded at different monitor elevations and probably in still air conditions so care must be taken when making comparisons between different makes and types. The quoted distances are likely to be reduced when the monitors are used under operational conditions. If at all possible, before purchase or operational use, this type of equipment should be operated at potential risk sites to ensure that acceptable throws and heights are achieved. This is especially true of risks involving storage tanks where the heights of the tanks and the large distance between the monitor (possibly positioned on or below a bund wall) and the tanks make the projection of foam into the tanks extremely difficult.

5.2.6 MX Hand-held Foam-making Branches

Medium expansion foam-making branches are generally designed to be used with SYNDET foam concentrates although other types, such as FP, AFFF, AFFF-AR, FFFP and FFFP-AR, may also be used. MX foam-making branches will produce foam at expansions usually ranging from 25:1 to 150:1. As a result of these higher expansion ratios,

Firefighting Foam – Technical **45**

Figure 5.8 A medium expansion hand-held foam making branch. (Photo: Mid and West Wales Fire Service)

5.2.7 LX and MX Hand-held Water Branch 'Snap-on' Attachments

'Snap-on' attachments are available for use with some hosereel and main line water branches which enable primary aspirated LX and MX foam to be produced. Generally, the foam produced by these attachments is not very well worked making it less stable (i.e. has much shorter drainage times) and less effective than that produced by purpose designed primary aspirating foam branches.

5.2.8 MX Foam Pourers

In addition to the MX hand-held foam-making branches, some free-standing MX foam pourers are also available. These are much larger than the hand-held models, have higher flow requirements and hence produce greater volumes of foam. However, as their name suggests, finished foam pours out of them rather than being projected. They have been designed to stand on their integral legs for the unattended delivery of MX foam into bunded areas, such as those surrounding fuel storage tanks. They operate in a similar way to the hand-held MX foam branches described above.

the projection distances of MX foam are much less than for LX foam. If at all possible, before purchase or operational use, this type of equipment should be operated to ensure that acceptable throws are achieved

With MX foam-making branches, an in-line inductor is generally used to introduce the foam concentrate as a premix. The branch then diffuses and aerates the stream of foam solution, and projects it through a gauze mesh to produce bubbles of a uniform size.

MX hand-held branches in use typically have nominal flow requirements ranging from 225 lpm to 450 lpm with inlet pressures ranging between 1.5 bar and 8 bar. The expansion ratio of the foam produced is usually claimed by the manufacturers to be in the region of 65:1 with throws ranging from 3 to 12 metres. Typical foam output is claimed to be approximately 13 m^3/min for 225 lpm branches and approximately 26 m^3/min for 450 lpm branches.

Typical models of MX foam pourers have nominal flow requirements of from 600 lpm to 1800 lpm when operated at 2.5 bar inlet pressure. The foam outputs of these are claimed to be approximately 24m^3/min and 72 m^3/min respectively at these operating conditions. This is at an expansion ratio of approximately 40:1.

5.2.9 HX Foam Generators

High expansion foam generators are designed to be used with SYNDET foam concentrate only and usually produce finished foams of expansion ratios of 200:1 to 1200:1.

Air is blown through the generator by a fan, foam solution is sprayed into the air stream, and this is directed onto the surface of a fine net screen. The air blowing through the net wetted with foam solution produces finished foam with a mass of bubbles of uniform size which, like the MX foam pourers, is "poured" rather than being "projected".

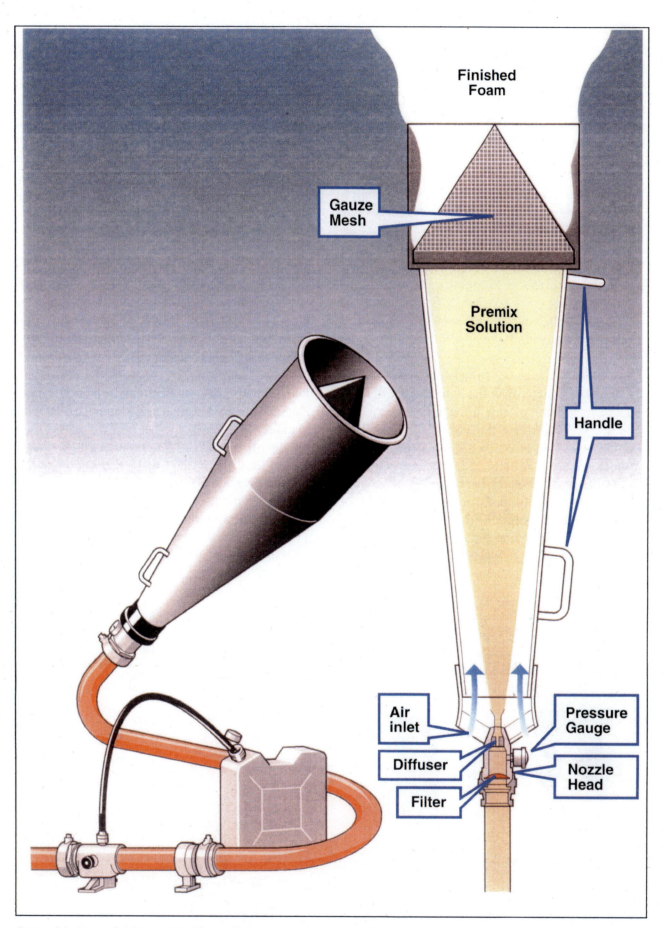

Figure 5.9 Principle of operation of a medium expansion foam branch pipe.

The generator fan may be powered by:

- a petrol engine;
- an electric motor;
- a water turbine which utilises the flowing foam solution immediately prior to it being sprayed into the generator.

The water turbine driven generators are obviously more suited to applications in areas where there is a flammable risk. Most HX foam generators can also be used as smoke extractors.

Figure 5.10 shows, in diagrammatic form, the essential principles of HX foam generators. Some generators require a separate in-line inductor but others are self-inducing and some are capable of being operated either way.

Some water turbine driven generators incorporate a 'by-pass' system. With the by-pass closed, all of the foam solution passing through the generator is used both for driving the turbine and for foam production. This produces a lower expansion HX finished foam containing a higher percentage of water. To overcome high back pressure, e.g. when forcing finished foam through long lengths of ducting or up to a height, the by-pass is opened, and some foam solution is thereby diverted to pass through the turbine to waste, giving less for foam production. This results in a higher expansion ratio, with the finished foam containing a lower percentage of water. It also slightly increases the water flow to the turbine, speeding up the fan and, consequently, the air flow.

Because the finished foam cannot be projected, it is often fed to the required application point through a large-diameter flexible tube or ducting. It can, however, be used without ducting, e.g. placed on the side of a ship's hold or in the doorway of an enclosure.

The larger HX foam generators are rather bulky items of equipment to carry on a first-line appliance, so they are usually brought by special vehicles when required. However, some lightweight generators have been developed that can fit into a standard appliance locker.

One typical large water turbine driven HX foam generator weighs 55 kg and is claimed by the manufacturer to produce at 7 bar inlet pressure, with a nominal flow of 210 lpm and the by-pass closed, 135 m^3/min of finished foam with an expansion ratio of between 500 and 700:1. At the same inlet pressure, but with a nominal flow of 225 lpm and the by-pass open, the foam output is claimed to be 155 m^3/min of finished foam with an expansion ratio of between 800 and 1200:1.

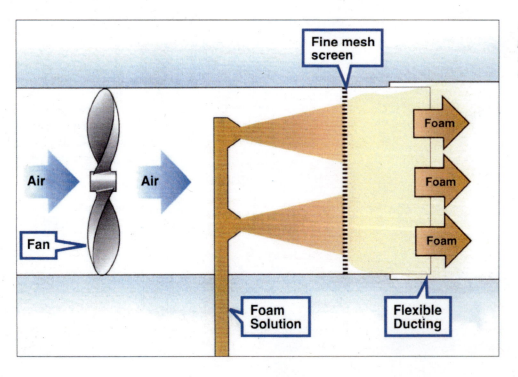

Figure 5.10 Essential principles of a High Expansion Foam Generator.

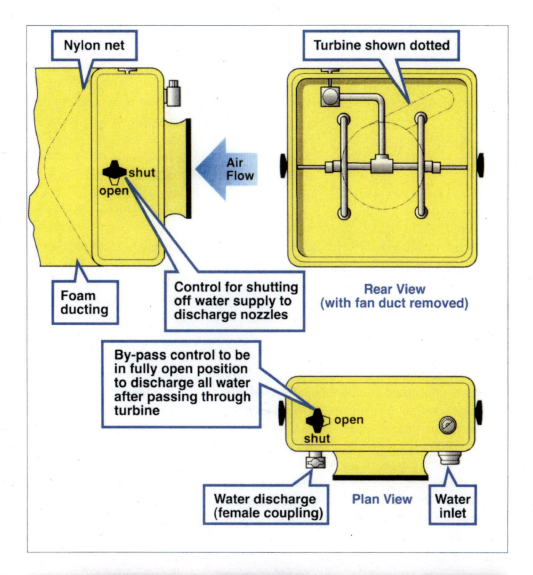

Figure 5.11 A typical high expansion foam generator.

Figure 5.12 Large HX foam generator stowed on foam tender (right). Large plastic bins (centre) are for decanting foam.
(Photo: Northern Ireland Fire Brigade)

Firefighting Foam – Technical 49

One typical small water turbine driven HX foam generator weighs 16 kg and is claimed by the manufacturer to produce at 7 bar inlet pressure, with a nominal flow of 245 lpm, 80 m^3/min of finished foam with an expansion ratio of 330:1.

5.3 Foam Concentrate Induction and Injection Equipment

5.3.1 General

Foam concentrate induction and injection equipment is used to introduce foam concentrate into the water supply in order to produce foam solution. There is a need for this equipment to work accurately in order to avoid wastage of foam concentrate and, more importantly, to help to ensure that the finished foam is of optimum quality.

The more concentrated a foam concentrate. (e.g. a 1% foam concentrate is more concentrated than a 3% foam concentrate) the lower the rate of flow that the foam concentrate is required to be introduced in to the water stream. Consequently, especially for 1% systems, even slight variations in the foam concentrate flow can result in much weaker/stronger foam solutions being produced than required. If too little foam concentrate is picked up, a weak foam solution will be formed which is likely to produce a poor blanket of quickly draining foam. If too much foam concentrate is picked up, a strong foam solution will be formed which is likely to produce foam that is too stiff to flow adequately across the surface of a fuel probably resulting in poor firefighting performance. In addition, expensive foam concentrate will be wasted and the possible overall duration of firefighting will be reduced due to rapid consumption of the available supply of foam concentrate.

Typically, variations in the accuracy of induction/injection equipment of + or - 10% of the required concentration are usually acceptable and are unlikely to affect firefighting performance, that is:

for 1% concentrate, induction rate to be between 0.9% and 1.1%;

for 3% concentrate, induction rate to be between 2.7% and 3.3%;

for 6% concentrate, induction rate to be between 5.4% and 6.6%.

Other levels of accuracy are often stated in standards and by equipment and foam concentrate manufacturers. Discussions with these organisa-

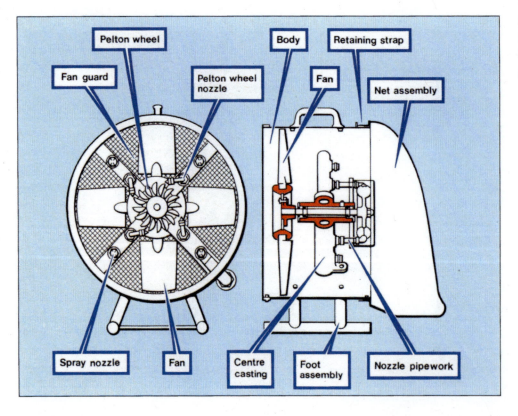

Figure 5.13 A typical small water-driven, high expansion foam generator.

tions may be necessary to ensure that these levels of accuracy are acceptable and will not affect the firefighting performance of the resulting foam or lead to unacceptable amounts of foam concentrate being wasted.

Some of the foam-making equipment described in the previous Section is self-inducing. In other words, the foam-making equipment can pick-up and mix foam concentrate with the water supply prior to producing finished foam. Generally, a pick-up tube, of a few metres length, is used to connect the foam-making equipment to a foam concentrate container. This method of induction is not always satisfactory for the following reasons:

- Control and operation of the induction system can be more carefully carried out at a safe distance from the fire.
- Movement of self-inducing foam-making equipment is restricted due to the need to be close to a supply of foam concentrate.
- Foam concentrate supplies have to be transported to the foam-making equipment.

For these reasons, foam concentrate is often introduced into the water supply line some distance away from the foam-making equipment. The types of induction equipment most commonly used by the fire service for this purpose are:

- in-line inductors;
- round-the-pump proportioners.

Whatever foam induction or injection equipment is used, its operation should be checked regularly to ensure that the rate at which the foam concentrate is introduced into the water stream is accurate. Such checks should involve the whole of the foam system to be used operationally, including the foam-making equipment, the foam concentrate, typical hose runs and typical pump/branch operating pressures and flows, to ensure that the system as a whole works as expected.

Problems that may occur include:

- long hose runs producing high back pressures which prevent the induction equipment proportioning correctly, or at all;
- foam concentrates that are too viscous to be picked up at the correct rate by the induction equipment. Different types and manufacturers versions of foam concentrates will be of different viscosity. These will affect the accuracy of the induction equipment.
- blocked or obstructed orifices within the induction equipment;
- poorly calibrated induction equipment (Note: the calibration of new induction equipment should always be checked with the foam-making equipment and foam concentrates it is to be used with);
- incorrect inductor for the foam-making equipment being used or for the required concentration of foam concentrate. (Note: some manufacturers colour code their induction and foam-making equipment to assist in identifying matched equipment)

This Section also includes information on methods that can be used to check the concentration of the foam solution that is produced by foam-making systems.

5.3.2 In-line inductors

An in-line inductor is placed in a line of delivery hose, usually not more than 60 metres away from the foam-making equipment. This allows the foam-making equipment to be moved around relatively freely without the additional need to move foam concentrate containers.

In-line inductors employ the venturi principle to induce the concentrate into the water stream. (Note: self-inducing foam-making branches also usually work in this way). Water is fed into the inlet of the inductor generally at a pressure of around 10 bar (see Figure 5.14). This passes through the smaller diameter nozzle within the inductor to a small induction chamber and then to on the inductor's large diameter outlet via a flow improver. As the water enters the small nozzle, its velocity increases dramatically causing its pressure to drop (the venturi principle) and the pressure in the induction chamber to fall below atmospheric pressure. This partial vacuum sucks the foam concentrate through the pick up tube and into the low pressure induction chamber.

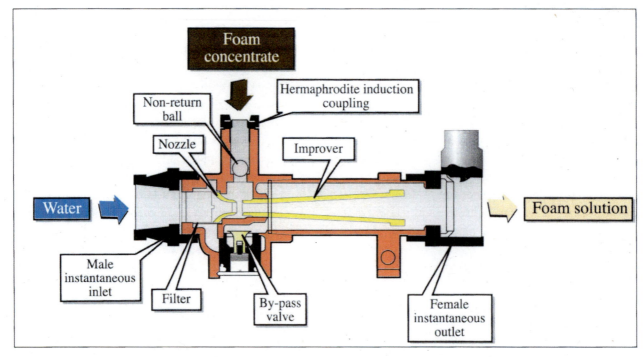

Figure 5.14 Principle of operation of an inline inductor.

A non-return valve (a ball is illustrated for this purpose in Figure 5.14) must be included in the foam concentrate pick up line to prevent water flowing back into the foam concentrate container.

There will always be a pressure drop across the inductor of at least 30% of the inlet pressure. This is necessary for the inductor to work properly. The pressure drop is due partly to turbulence and partly to the energy loss involved in the induction process. Pressure drops in excess of 70% have been recorded for hosereel in-line inductors (see this Chapter, Section 5.3.6).

If the back pressure at the outlet of the inductor is too high, this may result in the pressure drop across the inductor being less than required. In such circumstances, the velocity of the water travelling through the inductor nozzle would not be high enough to enable the pressure in the induction chamber to fall below atmospheric and so the inductor would fail to work. High back pressure can be caused by connecting too many lengths of hose between the outlet of the inductor and the foam-making equipment or through differences in elevation.

In order for it to operate effectively, it is important to match the pressure and flow characteristics of the foam-making equipment with that of the inductor. Inline inductors are usually identified by their nominal flow rate at 7 bar outlet pressure. Typical sizes of inductor are 225 lpm, 450 lpm and 900 lpm. Consequently, an inductor designed for a flow of 450 lpm can be used with one foam-making branch requiring 450 lpm or two foam-making branches, each requiring 225 lpm, and so on.

It is important to note however, that only one inductor should be used in any one hoseline. For instance, two 225 lpm inductors must not be used in a single hoseline to supply a 450 lpm foam-making branch. If this were to happen, the combination of the pressure losses across each of the inductors would result, at best, in the delivery to the foam-making branch of a very low pressure and low flow foam solution of incorrect concentration.

Some inductors contain a bypass valve (see Figure 5.14) which assists in enabling them to maintain induction over a range of inductor inlet pressures, often 4 to 10 bar, when using the correct foam-making equipment. In addition, the bypass valve can help to minimise the pressure drop across the inductor and assist in overcoming some slight mismatch problems caused by using inductors and foam-making equipment of similar nominal flowrates but from different manufacturers. However, the accuracy of inductors containing bypass valves can vary considerably with pressure although they will tend to be slightly more accurate at varying pressures than those without bypass valves (see below).

Other inductors, without the bypass, will only give the correct induction rate at one particular inlet pressure and flow, e.g. at 7 bar and 225 lpm. Operation other than at the pressure and flows recommended by the manufacture will result in inaccurate foam pick up rates, or no foam pick up at all.

Fixed and variable rate in-line inductors are available. Fixed rate inductors can only be used at one induction rate, generally either 1%, 3% or 6%. The induction rate of variable in-line inductors can usually be varied anywhere between 1% and 6% by the use of a control knob (Figure 5.15).

Practically all in-line inductors are designed to induce the foam concentrate through a pick-up tube placed in a drum or similar container. They can, however, also be used in conjunction with a pressurised foam concentrate supply (Figure 5.16).

The advantages of the use of in-line inductors are:

- generally the cheapest induction system available;
- simple, robust and with few moving parts;
- quick deployment/redeployment on the fire ground;
- foam solution does not pass through the pump or appliance pipework making clean up easier and reducing the possible corrosive effects of the foam solutions.

The disadvantages of in-line inductors are:

- for optimum performance, the inductor must be matched to the foam-making equipment;
- for optimum performance, the inductor must be matched to the type and concentration of foam concentrate in use;
- pressure losses through the inductor in excess of 30% can be expected at the normal working pressure range when using matched foam-making equipment;
- accuracy of proportioning will vary with pressure.

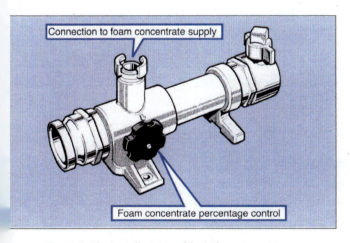

Figure 5.15 An inline variable inductor.

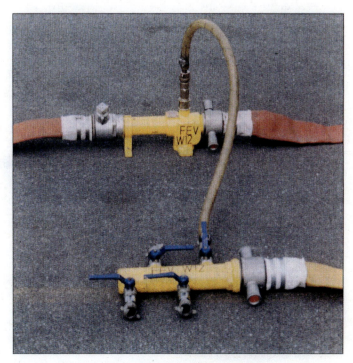

Figure 5.16 An inline inductor (top) connected to a pressurised foam concentrate supply (bottom).
(Photo: Mid and West Wales)

Firefighting Foam – Technical 53

5.3.3 Round-the-pump Proportioners

This type of inductor is connected across a pump and can either be a permanent fixture in the appliance or, with adapters and connecting hoses, stand alone. Two typical available models are one with a nominal induction flow range of 0–45 lpm and the other with a nominal induction flow range of 0–90 lpm. The induction flow can be altered within these ranges by the use of a rotating grip handle on the body which has a scale calibrated in litres per minute.

Figure 5.17 shows a typical round-the-pump proportioning system where an appliance has a built-in foam concentrate tank. When pumping begins, some water flows to the deliveries and some passes to the proportioner. The proportioner induces foam concentrate to produce a rich foam solution which passes back to the suction side of the pump. Before re-entering the pump, the foam solution mixes with a fresh intake of water, and is consequently diluted to the required concentration. Most of it then passes to the deliveries, while a small amount returns to the proportioner where more concentrate is induced, and the sequence is repeated.

Isolating valves can be incorporated to cut off the system when foam is not required. Various other valves in this sort of system are incorporated to:

- drain the foam concentrate tank;
- flush the system;
- connect a pick-up tube in case a foam supply other than that contained in the appliance foam concentrate tank needs to be used.

Although this proportioner has an operating pressure range of between 3 and 14 bar, the recommended pressure is 7 bar, with a water requirement of 193 lpm.

The induction rate for a round-the-pump inductor has to be selected by a dial calibrated in litres per minute. Consequently, the operator must know the flow rate at which the foam equipment is operating in order to be able to calculate the correct pick up flow rate for the concentration of foam concentrate being used. For instance, if the supply to the foam-making equipment is 193 lpm, and 3% concentrate is being used, then the inductor dial should be set as follows:

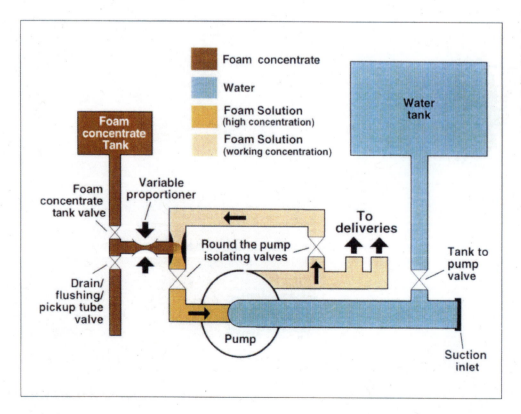

Figure 5.17 Diagrammatic layout of a round-the-pump proportioner system where there is a built-in foam tank.

| Dial setting | = | Foam equipment flow rate | × | Required percentage concentration / 100 |

= 193 × 3/100

Dial setting = 5.8 lpm

However, each time the flow to the foam-making equipment changes, then a new setting would have to be calculated to maintain accurate foam concentration.

Round-the-pump systems require the pressure on the suction side of the pump to be less than one-third of the pressure on the delivery side in order to function correctly. If this is not the case, then water may be forced into the foam concentrate container. Pressure control valves are available which can help to reduce this problem (see below).

Another drawback of this type of equipment is that the foam solution has to pass through the pump casing. This may cause corrosion problems within the pump and other areas of the appliance where the foam solution may enter, such as the water tank and pipework. In addition, the orifices within the inductor are extremely small; these can easily be blocked by small pieces of debris or foam concentrate sludge. Once blocked, the system must be taken apart and the debris removed. For systems fixed on to appliances, this requires the appliance to be taken off the run.

However, one of the major advantages of round-the-pump proportioners is that their use does not result in pressure losses at the output side of the pump.

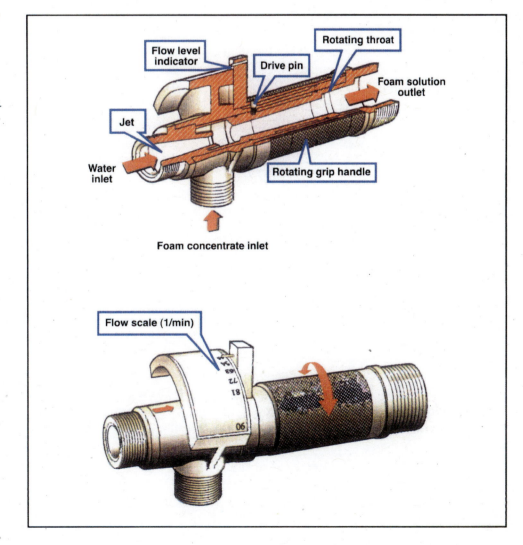

Figure 5.18 A typical round-the-pump proportioner.

Top: cutaway view.

Bottom: external view.

Firefighting Foam – Technical 55

To summarise, the advantages of round-the-pump proportioners are:

- relatively inexpensive, although more expensive than in-line inductors;
- they can provide a variable, accurate induction flow rate over a wide range;
- can be used as a fixed or temporary system on appliances;
- wide operating pressure and flow range;
- do not cause the pressure drops in the delivery hose that are associated with in-line inductors, this allows foam solution to be supplied through extended lengths of hose.

The disadvantages are:

- to maintain accurate concentration of foam concentrate, the operator must continually calculate and adjust the foam concentrate flow rate;
- foam solution is passed through the pump. There is concern regarding corrosion of the pump and other associated areas; thorough flushing after use is essential.
- where the pump feeds more than one branch, there is a need to match the pump output and the concentrate flow to take account of the number of branches in use at any one time.
- pressure control valves are needed where water feed into the suction side of the pump is high (see below)
- there are very small orifices within the system which can easily be blocked by debris or foam concentrate sludge.

5.3.4 Pressure Control Valves

A round-the-pump proportioner will only function correctly if the pressure on the suction side of the pump is less than one-third of the pressure on the delivery side. If this limit is exceeded, when pumping from a hydrant for instance, the back pressure acting on the outlet of the proportioner will be sufficient to inhibit the induction of foam concentrate. In extreme conditions, where no non-return valves are present, water may feed back into the foam concentrate container.

To prevent this situation from arising, a pressure control valve may be used with the proportioner. The valve reduces the pressure in the pump inlet line to one-fifth of the incoming pressure, thus bringing it within the required limit under most of the operational conditions that are likely to be encountered. The valve may be fitted as an integral part of the pipework system on an appliance, or used as a portable unit inserted into the pump inlet line at any convenient position.

Figure 5.19 illustrates a typical pressure control valve. Water, under pressure from the hydrant, passes through the valve over a movable butterfly. This butterfly is connected to a hydraulic piston which receives pressure from both sides of the butterfly. The area of the piston which is subjected to pressure on the upstream side is one-fifth of the area of the piston on the downstream side, so the forces acting on the piston will balance when the downstream pressure is one-fifth of the upstream pressure.

If the upstream (i.e. hydrant) pressure increases, the downstream side will experience a proportionally greater pressure increase. This will immediately cause the piston to move, closing the butterfly and reducing the flow through the valve, thereby reducing the downstream pressure until the 5:1 ratio is restored. If the hydrant pressure falls, the reverse process will occur.

5.3.5 Pressurised Foam Supply

(a) General

At a large incident requiring perhaps several large foam monitors, bulk supplies of foam concentrate will be required. In these circumstances, the conventional system of inducing the concentrate via a pick-up tube may be impractical, for the following reasons:

- It may not be feasible to use foam concentrate drums to supply the inductor because of the frequency with which they would have to be refilled or replaced.
- The use of open-topped portable dams may not be entirely satisfactory because when using some systems the foam concentrate tends to aerate and this can interrupt the

Figure 5.19 A typical pressure control valve with cutaway drawing and schematic diagram.

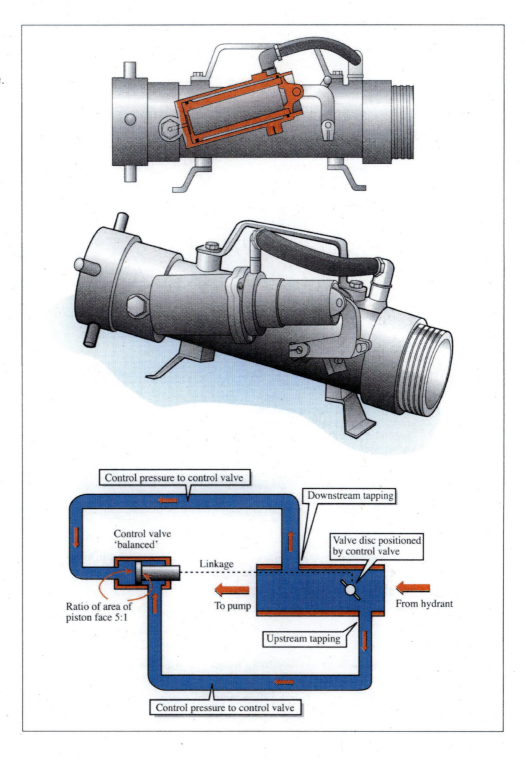

supply. Spillages can also occur and the logistics of keeping the dams topped up need to be considered. Debris may also enter the dams which may lead to blockages of the induction system.

- Since the pick-up system requires the concentrate container to be positioned very near to the inductor, it may not be possible for a bulk supply to be positioned close enough to supply the inductor directly.

Even at smaller incidents, where it is practicable to use foam concentrate drums, it may be impossible to determine when the concentrate is about to run out. This could lead to water being discharged onto the fire. This may also occur whilst the pick-up tube is being transferred when a container becomes empty.

Some firefighters adapt foam concentrate containers by cutting off the top so that the contents

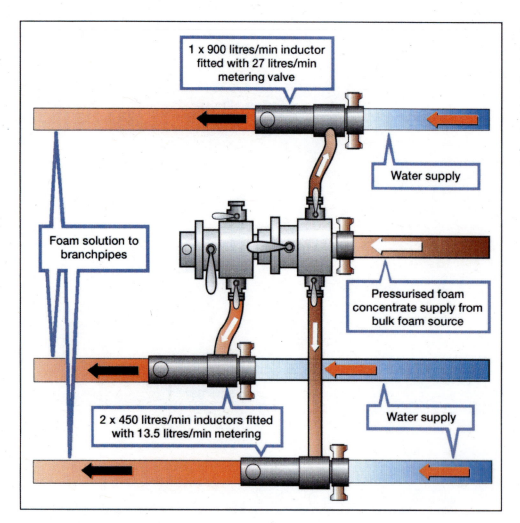

Figure 5.20
Diagram showing one 900 litre/min and two 450 litre/min foam branchpipes receiving a pressurised foam concentrate supply from a foam tanker or foam main.

can be seen and topping up is made easier. However, this should be done with care because 'swarf' produced when cutting off the tops can be picked up and can cause blockages of the induction system.

To overcome all of the above difficulties, many brigades have developed pressurised foam concentrate supply systems in which the foam concentrate is pumped from bulk storage containers directly to the delivery equipment. This is often achieved by utilising the pumping units on foam tenders to convey the foam concentrate to the induction device, which may take the form of an in-line inductor or a constant flow valve.

Brigades have their different versions of this system, but they will all usually include some type of:

- distribution manifold, and
- metering device.

When pumping foam concentrate to in-line inductors in particular, care should be taken to ensure that the system has been correctly designed for this situation. This is mainly because these inductors are calibrated for their normal operating mode where they create their own small partial vacuum in order to suck up foam concentrate (see this Chapter, Section 5.3.2). However, when foam concentrate is pumped under pressure directly to them, this will act in addition to the partial vacuum and will result in foam concentrate being introduced into the system at a much higher concentration than required.

Three other methods of feeding foam concentrate under pressure into hose lines without the use of in-line inductors are also briefly discussed below, these are:

- pelton wheel in-line foam injection;
- pre-induction units;
- direct coupled water pump.

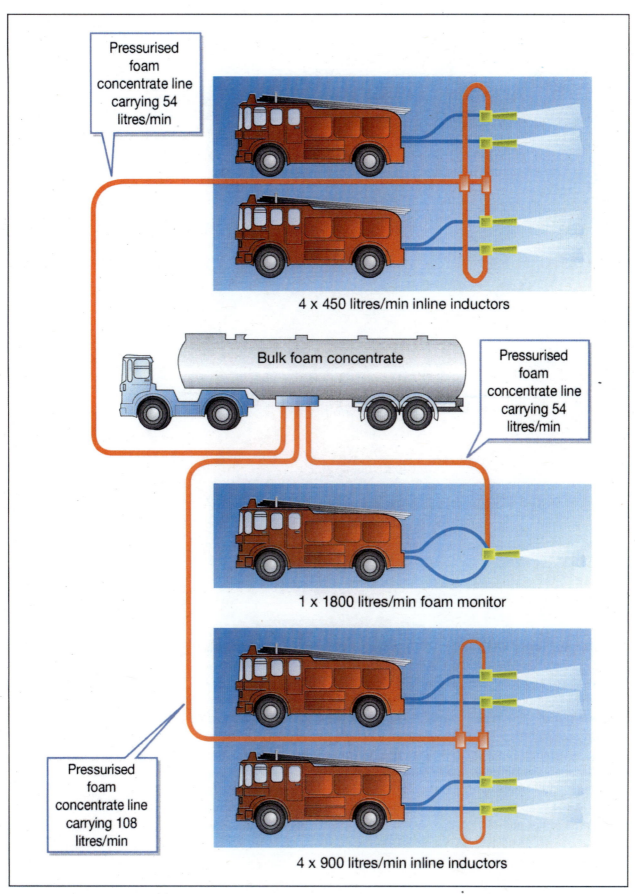

Figure 5.21 Diagrammatic layout of an incident requiring large quantities of foam concentrate supplied from a bulk foam carrier.

Firefighting Foam – Technical

Figure 5.22
Pressurised foam
concentrate supply
being got to work.
(Photo: Surrey Fire and Rescue Service)

(b)   Distribution Manifold

Various designs have been devised by brigades, some incorporating a metering device. Figure 5.23 shows a typical distribution manifold which consists of a standard male instantaneous coupling leading to a manifold having two controlled outlets with 20 mm hermaphrodite couplings, one on each side, and a full-bore on/off valve. The manifold finally has a standard female instantaneous coupling at the other end.

This type of manifold is capable of feeding one or two in-line inductors through 20 mm hose, each line passing up to 70 litres of foam concentrate per minute. If more than two in-line inductors need to be used, a second manifold can be added to the first one, either directly or via additional lengths of hose. The shut-off valves on the manifolds are opened or closed according to the number of in-line inductors to be supplied.

(c)   Metering Devices

In order to ensure the optimum output of foam-making equipment, the correct amount of foam concentrate should be fed to the inductors at all times. To ensure this, a metering device, or constant flow valve as it is also known, is inserted into the line. There are many of these types of devices available.

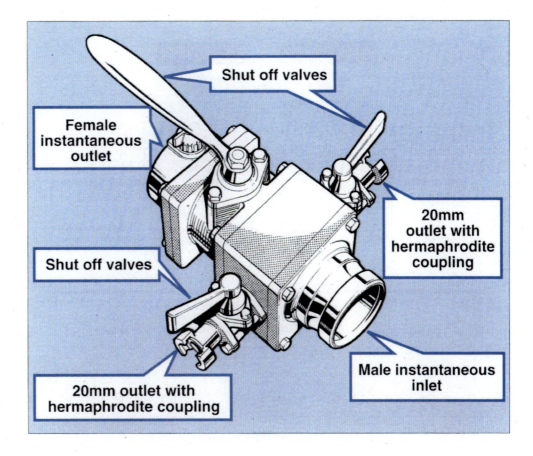

Figure 5.23 A typical distribution manifold.

One example of a metering device consists of a flexible ring resting on a tapered seating. At low pressure the ring is fully expanded, giving the maximum orifice opening. As the pressure rises, the ring is compressed and forced gradually down the tapered seating, progressively decreasing the size of the orifice. The combination of increased pressure and decreased orifice size maintains a constant flow (Figure 5.24).

In another example (see Figure 5.25), a neoprene diaphragm (shaped like a plug) is located above a profiled orifice. When subjected to pressure variations between 1 and 14 bar, this diaphragm flexes onto the orifice, thereby increasing or decreasing the available orifice area and maintaining a constant rate of flow.

These valves may be inserted at the inlet to each in-line inductor or at some other point in the foam concentrate delivery line. Several brigades have had foam-making equipment modified so that the valve is incorporated within them. It is, of course, essential that a metering device of the correct flow rating for the equipment is used.

(d) Inline Foam Injection (Pelton Wheel)

As mentioned earlier, high pressure losses, in excess of 30%, can be expected when using in-line venturi inductors. It is not unusual for this loss together with hose, monitor and nozzle losses to add up to a total pressure loss that makes the performance of some pieces of foam-making equipment ineffective, particularly in terms of throw.

One alternative system is to make use of a pelton wheel driven positive displacement pump which will introduce foam, from a storage tank or foam dam, into the delivery hoses through a regulating valve. This valve can be adjusted to suit the injection rate required and once set will inject at the required percentage regardless of pressure fluctuations in the delivery hose. The units can be supplied with either fixed or adjustable induction rates to suit the circumstances.

(e) Pre-induction Units

This system employs two induction units. A pre-induction unit is installed near a hydrant and draws

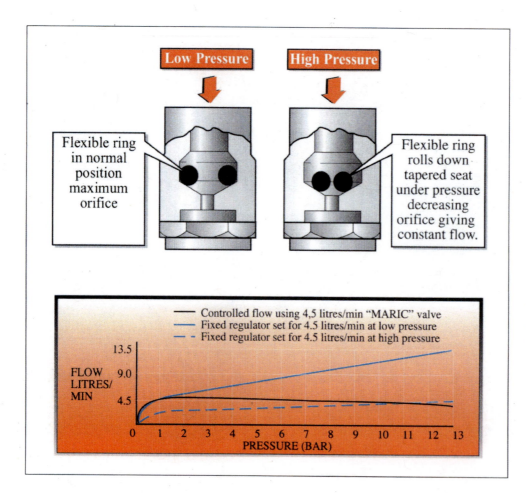

Figure 5.24
Principle of operation of one type of metering device (Maric valve).

Graph showing the performance of a 4.5 litre/min Maric valve.

concentrate from a reservoir to produce a concentrate rich solution, generally in the region of one part water to two parts foam concentrate. This is fed to a specially designed self-inducing foam-making branch. It is at the branch where the required concentration is achieved.

By using two stages of induction and making use of the pressure and flow from a separate hydrant, a much lower pressure loss is experienced across the inductor in the branch. However, the pre-induction unit must be accurately matched to the foam-making branch.

When using 75mm diameter hose and large monitors, the distance from a pre-induction unit to the monitor can be in excess of 750m.

(f) Direct Coupled Water Pump

These usually consist of two positive displacement rotary pumps, a water motor and a foam concentrate motor, that are linked via a common shaft. The water motor is connected into the main water feed line to the foam-making equipment. As water passes through this motor, it turns and drives the foam concentrate motor which injects foam concentrate into the water stream discharging from the outlet of the water pump. The capacities of the motors are carefully chosen so that the correct percentage of foam concentrate is injected into the water line. Due to the linkage between the motors, the percentage concentration remains correct over a wide range of flows through the water motor.

Typical portable versions cover various flow ranges between 200 and 2000 litres per minute at a maximum pressure of 15 bar. The induction rate is usually either fixed at 3% or is adjustable from 3% to 6%.

Disadvantages of this type of system are that they are expensive and that they can produce pressure drops of 25 to 30% of the inlet pressure.

Figure 5.25 Principle of operation of the 'Mobrey constaflo' valve.

(1) Single valve.

(2) Multiple valve.

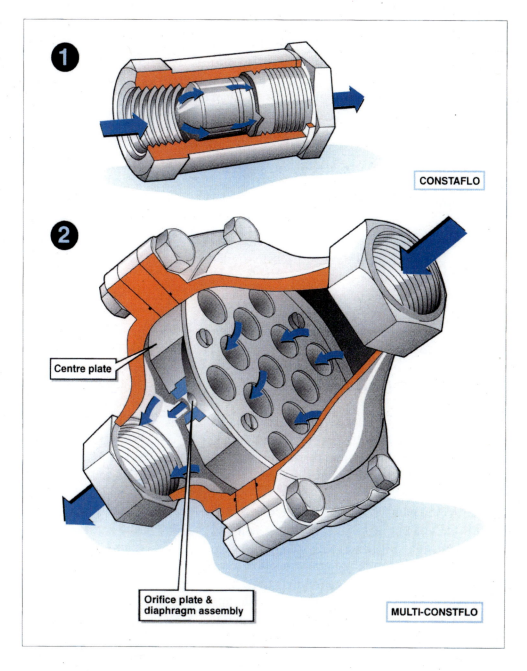

5.3.6 Hosereel Foam Induction and Injection Systems

(a) General

There are four categories of system most often available for use by brigades for the induction of foam concentrates into high pressure hosereels (Reference 8). They are:

- Premix systems
- Round-the-pump systems
- Injection into pump inlet
- In-line inductor

The basic principles of operation of these systems are given below. This is followed by suggestions for an operational requirement for a system to induce additives into the high pressure hosereels of a first-line appliance. This operational requirement is then briefly compared with the typical performance of some existing hosereel induction systems.

(b) Premix

Premix systems involve foam concentrate being mixed, to the correct concentration, with, for instance the whole contents of an appliance water tank. A true premix system is ready-mixed in

Firefighting Foam – Technical 63

advance of use while a dump tank premix system drops foam concentrate into the appliance water tank only when required. For a dump tank system, the whole contents of the tank become a premix and there may be a significant wastage in foam concentrate if this is not completely used at an incident. Conversely, for a true premix system, there may be problems in maintaining the correct foam solution concentration when 'topping-up' the tank.

(c) Round-the-pump

A typical round-the-pump system is described earlier in this Chapter in Section 5.3. Other similar systems are available for use in hosereel systems, some of which use flowmeters, valves and microprocessor control. These match the foam concentrate flow rate to the water flow rate to maintain the required foam concentration.

(d) Injection in to Pump Inlet

Injection in to pump inlet systems, as the name suggests, involves the injection of foam concentrate in to the eye of the pump. Injection is usually either by electric pump or by gravity feed.

For an electric pump system, a regulator is used to control the amount of foam concentrate that can enter the high pressure side of the pump. Consequently, this can be calibrated to allow various concentrations of foam concentrate to be used through one or two hosereels (or a main line branch if necessary). In the gravity fed systems, precisely sized orifices are used to regulate the supply of foam concentrate into the pump. Several orifices of different sizes can be included in the system. These also allow different concentrations of foam concentrate to be used or for one or two hosereel branches (or a main branch) to be used in conjunction with the system.

(e) In-line Inductors

Hosereel in-line inductors work in the same way as the main delivery types previously described in Section 5.3 of this Chapter.

However, tests carried out on one hosereel in-line inductor (Reference 8) found that there was a pressure loss in excess of 50% across the inductor when the recommended jet/spray branch was operated on jet. When the branch was operated on spray, the pressure loss across the inductor was in excess of 70% and, when the effects of the hosereel tubing were taken in to account, the total pressure loss was in excess of 90%. Consequently, using this in-line inductor at the appliance pump, connected to the branch with 3 lengths of 19mm hosereel tubing and at a pump pressure of 26 bar, resulted in a branch pressure of less than 3 bar.

(f) Suggestions for an Operational Requirement for a Hosereel Induction System

The following are suggestions for inclusion in an operational requirement for a system to induce all types of foam concentrates into the high pressure hosereels of a first-line appliance:

- It should be capable of inducing all types of foam concentrate at selected concentrations within the range 1% to 6%. For alcohol resistant foams, it must be possible to select 3% concentration for hydrocarbon fires and 6% concentration for alcohol fires.

- The accuracy of induction should be maintained over the varying flow and pressure conditions from one or two hosereels up to a total flowrate of 300 lpm.

- The induction system should be accurate to plus or minus 10% of the correct concentration, that is:

    for 1% concentrate, induction rate to be between 0.9% and 1.1%;
    for 3% concentrate, induction rate to be between 2.7% and 3.3%;
    for 6% concentrate, induction rate to be between 5.4% and 6.6%.

- When foam concentrate is required for the hosereels only, ideally, no foam solution should be available from the main deliveries.

- When the hosereels are off, the foam concentrate flow should be zero. If foam concentrate is not required during use then it should be possible to turn the supply off. No water should flow into the foam concentrate container at any time.

- The system must work when pumping from the appliance water tank, a pressure fed supply or open water.

- The system should not adversely affect branch performance due to, for instance, high pressure losses.

- The system should be capable of continuous operation especially while the foam concentrate supply is replenished.

- It should be possible to retrofit the system to appliances.

In addition to these, the desirability of having foam solution passing through the appliance pump is also an important factor to be considered.

It is unlikely that any hosereel induction system will meet all aspects of this suggested operational requirement. The only system at present that does not pass foam solution through the pump is the in-line inductor. However, the use of this results in pressure drops in excess of 70% which would adversely affect branch performance.

Systems are often unable to maintain correct rates of foam proportioning over the range of flows likely to be experienced on the fireground. All have to be operated in very restricted ranges of flow and pressure in order to maintain accurate induction rates.

The outline of the operational requirement given above is a good starting point for brigades to evaluate any new hosereel induction system that may come on to the market.

5.4 Compressed Air Foam Systems (CAFS)

Compressed Air Foam Systems (CAFS) are designed to produce aspirated finished foams without the need of a foam-making branch. CAFS consist of a water pump, a foam concentrate injection pump and an air compressor which combine to produce an aerated foam at the delivery of the pump. CAFS can be appliance or trailer mounted and can be supplied with a range of water pumps, concentrate injection pumps and different air compressors depending on the requirements of the user.

It is claimed by the manufacturers of these systems that they have longer throws than conventional fire service equipment and that they produce better worked foam with expansion ratios adjustable between 7:1 and 30:1. When used with class A foam concentrates, it is claimed that the resulting foam will stick to vertical surfaces and remain there for long periods of time. This is said to cool and insulate the material and to prevent the spread of fire from radiated heat. CAFS may also be used with other types of foam concentrates.

Tests (Reference 9) have shown that a CAFS can throw foam further than conventional UK fire service foam-making branches whilst producing a well worked low expansion foam. The system also produced a medium expansion foam with FP that was very sticky and could be used to coat vertical surfaces.

5.5 Methods For Checking Foam Solution Concentration as Produced by Foam-making Equipment

5.5.1 General

It is important that the whole foam-making system is regularly checked to ensure that it works as expected and that the concentration of the foam solution produced is as required. The following two methods can be used to check the concentration of foam solution:

- The use of a refractometer.
- Foam concentrate and water flow measurements

On first inspection, it would appear that the refractometer method is perhaps too difficult to use. However, once some experience has been gained in its use, the refractometer method will prove more accurate and simpler to use than attempting to measure liquid flows, especially if relatively accurate flowmeters are not available.

5.5.2 Refractometer Method

One method of checking the concentration of the foam solution is by the use of a refractometer. When used for this application, a refractometer measures the change that occurs in the direction of travel of light at the junction of foam solution with glass in terms of its refractive index. There is a straight line relationship between refractive index and solution concentration.

A refractometer looks similar to a small telescope with an eyepiece at one end and a hinged prism box at the other. They are available from laboratory suppliers and are relatively easy to use with care. They are widely used in manufacturing industries for measurements of concentrations of fruit juices, battery acids, wines, soft drinks, starches, glues and so on.

The procedure for using a refractometer is as follows; a calibration curve should be produced for the foam concentrate under test. Ideally, this should be produced prior to each occasion that the refractometer is used. It is important that the actual foam concentrate that will be passed through the system, and water from the same supply, be used to make up samples of various concentrations of foam solution.

At least 3 calibration points should be chosen which cover the range of from 0.5 times to 2 times the expected inductor pick-up concentration. For instance, for a system supplying 3% concentrate as a 3% foam solution, the calibration samples should be 1.5ml, 3ml and 6ml respectively of 3% foam concentrate made up and thoroughly mixed with water each to make 100ml of foam solution (i.e. these would produce 1.5%, 3% and 6% foam solution samples). The refractive index of each of these should be measured using the refractometer and a graph plotted of refractive index against percentage concentration. All of the calibration points should then be joined with a straight line.

The foam-making system should then be run up to its normal operating conditions. A sample of foam solution should be collected from the foam-making equipment 30 seconds after foam production commences. This may mean the collection of foam in a large, clean bucket, with the foam solution that drains off being used. The refractive index of the collected foam solution should then be measured using the refractometer and its concentration read off from the calibration graph.

Different foam concentrate types produce foam solutions with different refractive indices and refractometers only cover limited ranges of refractive index. Consequently, care must be taken to choose the correct refractometer to cover the expected range of the refractive indices of the foam solutions to be tested.

5.5.3 Flow Method

Another method of checking the induction rate is to use a wide top container for the foam concentrate, such as a bucket, with calibrated marks perhaps every five litres. The amount signified by each mark will depend on the rate of foam concentrate pick-up expected from the induction equipment and the size of the container. Ideally, the container should contain the foam concentrate that will be used operationally with the foam equipment. Once the foam equipment has been run up to the required operating conditions, the pick up tube should be inserted into the container. The time taken for the level of the concentrate to fall by, for instance, 5, 10, or 15 litres, should then be measured. The induction rate can be calculated as follows:

Flow rate of foam concentrate (lpm)

$$= \frac{\text{Amount of foam concentrate used (litres)} \times 60}{\text{Time taken to use it (seconds)}}$$

Induction rate of inductor (percent)

$$= \frac{\text{Flow rate of foam concentrate (lpm)}}{\text{Flow rate through foam-making equipment (lpm)}}$$

The flow rate through the foam-making equipment should, if possible, be measured with a flowmeter. If a flowmeter is not available then the flow rate information provided by the manufacturer will have to be used although this may not be particularly accurate for normal fire service operating conditions.

The following is an example of the use of the above calculation. A LX foam-making branch operates, with an in-line inductor, at 225 litres per minute at a branch pressure of 7 bar. The inductor was timed to pick up 5 litres of foam concentrate in 45 seconds.

The induction rate of the inductor is calculated as follows:

Flow rate of foam concentrate

$$= \frac{5 \times 60}{45}$$

$$= 6.7 \text{ lpm}$$

Induction rate of inductor

$$= \frac{6.7}{225} \times 100$$

$$= 3.0\%$$

Firefighting Foam – Technical

Chapter 6

Chapter 6 – Categories of Fire and the use of Firefighting Foams against them

6.1 Classes of Fire

In the UK the standard classification of fire types is defined in BS EN 2: 1992 as follows:

Class A: fires involving solid materials, usually of an organic nature, in which combustion normally takes place with the formation of glowing embers.

Class B: fires involving liquids or liquefiable solids.

Class C: fires involving gases.

Class D: fires involving metals.

Electrical fires are not included in this system of classification (see this Chapter, Section 6.2).

In the following Sections, the general principles of extinguishment, particularly in relation to firefighting foams, are reviewed for each of the above classes of fire.

6.1.1 Class A fires

Class A fires are those which involve solid materials usually of an organic nature such as wood, cloth, paper, rubber and many plastics.

Some manufacturers of AFFF, AFFF-AR, FFFP, FFFP-AR and SYNDET foams state that their products may be used as wetting agents at between 0.1% and 3% concentration to assist in the extinction of class A fires. For these fires, AFFF, AFFF-AR, FFFP and FFFP-AR may be used at low and medium expansion while SYNDET foams may be used at low, medium or high expansion.

There are said to be advantages in the use of wetting agents when fires become deep seated. In these conditions, water can be slow to penetrate. A wetting agent that reduces the surface tension of the water is claimed to greatly improve penetration to the seat of these types of fire. When a wetting agent is employed, a deep seated fire is predominantly extinguished by the cooling effect of the water mix rather than by the smothering effect of any foam that may be produced.

Surfactant based foams display some wetting agent properties, but are more expensive than products sold purely for their wetting agent characteristics. From time to time, a few brigades take advantage of these wetting agent properties by using AFFF not only for class B fires (see below), but also, they claim, to make better use of limited water supplies on Class A fires. It is claimed that the increased cost in agent is often justified by reduced water damage to the property.

Tests have indicated that in some circumstances the addition of some foam concentrates to water can help in reducing the severity of a Class A fire when compared to the use of water alone (Reference 10). In particular, when applied by spray to wooden crib fires, secondary aspirated AFFF, and to a slightly lesser extent, FFFP, AFFF-AR and SYNDET, performed significantly better than water. Several wetting agents were also tested but they did not perform much better than water. These results seem to indicate that wetting properties may not alone quickly and effectively deal with Class A fires involving wood. The smothering characteristics of the foams may also be helping. (In fact, this is the principle under which American 'Class A' foams have been developed – see Chapter 2, Section 2.1.7.)

During these tests, because of the size and shape of the fires, some areas of the cribs were not adequately reached by the spray. Consequently, tests were also performed using jet applications of water, primary aspirated AFFF and secondary aspirated AFFF. There was little difference in the firefighting performances of these indicating that if adequate amounts of water can be applied to all areas of a wood fire, it will perform just as well as a primary aspirated or secondary aspirated foam when used in the same conditions.

Medium and high expansion foam have been advocated for indoor use on class A fires. The confinement provided by the walls of buildings allows the foam to accumulate into a thick blanket and also protects the foam from being torn apart by the wind. The mechanism put forward for extinguishment is that the foam cuts down the movement of air which supports combustion. There is a cooling effect as water from the foam evaporates, and the steam generated will also tend to reduce the oxygen level in the air surrounding the fire. If the foam blanket is deep enough, it will exert enough downward pressure to enable it to refill holes opened up when the foam is destroyed by the heat from the fire. Materials and structural members that would otherwise be exposed are shielded from heat radiation by the foam.

Although high expansion foam can be effective, the main practical drawback is that firefighters cannot be sure that the fire has been extinguished. It can be dangerous to enter a deep foam blanket to track down the seat of a fire since there is a chance of sudden exposure to heat and products of combustion. Under some conditions, the fire can continue to burn for a considerable period at a reduced rate supported by the air released from the foam as it breaks down.

The use of medium expansion foam against indoor class A fires, such as in warehouses, could be a more effective and efficient use of foam. It should be possible to restrict the foam application so that the area of origin of the fire is kept under observation whilst maintaining sufficient foam flow to force the foam onto the fire.

6.1.2 Class B Fires

(a) General

Class B fires are those which involve flammable liquids, liquefiable solids, oils, greases, tars, oil based paints and lacquers (i.e. flammable and combustible liquids). Combustion of these occurs entirely in the vapour that is present above the surface of the liquid. For firefighting purposes, Class B liquids can be subdivided into three categories, each requiring different properties from firefighting foams in order to achieve effective and efficient fire control and extinction.

The categories are:

- high flash point water-immiscible Class B liquids;
- low flash point water-immiscible Class B liquids;
- water-miscible Class B liquids.

Some high flash point liquid hydrocarbon fires, such as those involving fuel oils, can, under very controlled conditions, be extinguished using only the cooling effect of water.

However, most low flash point hydrocarbon fires, such as those involving petrol, cannot be extinguished by water alone as the fuel cannot be lowered to a temperature where the quantity of vapour produced is too small to sustain burning. In addition, water is generally much denser than liquid hydrocarbons, consequently, when applied during firefighting, it immediately sinks below their surfaces without having any beneficial effect, in firefighting terms, on the fire. In fact, the application of water may cause the surface area of the fire to increase and spread to previously unaffected areas.

Foam is generally applied to both high and low flash point hydrocarbon fuel fires because it provides a visible blanket which controls and extinguishes these fires faster and more effectively than water.

The three categories of Class B liquids and their firefighting characteristics are described in the following Sections.

(b) High Flash Point Water-immiscible Class B Liquids

Water-immiscible liquids with high flash points, or class C petroleum liquids, are those with a flash point above 55°C such as gas oils, some diesel oils, heavy fuel oils and heavy lubricating oils. At normal ambient temperatures these liquids have low vapour pressures and so do not generate flammable concentrations of vapour.

Water spray can be used to extinguish fires in high flash point liquids since the cooling effect of water is sufficient to reduce the generation of vapour to below the concentration needed to sustain combustion.
Firefighting foams are very effective against this type of fire giving very rapid control and security against reignition, however, use of water spray can be perfectly satisfactory and far less expensive in many cases.

The primary mechanisms by which foams extinguish high flash point liquid fires is by cooling the liquid surface and cutting out back radiation from the flames. The smothering action of foam plays a relatively insignificant role.

(c) Low Flash Point Water-immiscible Class B Liquids

Water-immiscible liquids with low flash points, or class A and B petroleum liquids, have flash points below 21°C and 55°C respectively. These include class A petroleum liquids such as aviation gasoline, benzene, crude oil, hexane, toluene and petrol (including lead-free), and class B petroleum liquids such as avtur jet fuel and white spirit.

Spills or pools of low flash point liquids can produce flammable vapour under normal ambient temperatures, and flammable or explosive concentrations can accumulate at low level, since most of the vapour will be heavier than air.

Water sprays are unsuccessful in extinguishing fires in low flash point liquids because vapour generation is not sufficiently reduced by the degree of cooling achieved. However, considerable reductions in flame height and radiation intensity can be achieved with water spray application. Obviously, care should be taken to ensure that the fuel does not overflow any containment. In addition, where the fuel is not contained, the application of water will result in further fuel spread.

Firefighting foams are effective on low flash point liquids because they trap the vapour at, or just above, the liquid surface. The trapped vapour then sets up an equilibrium with the liquid which prevents further vapour generation. Where deep foam blankets can be formed, such as in storage tanks with a large freeboard, this process may be assisted by the increased pressure exerted by the heavier blanket. Film-forming foams produce a thin film on the surface of some of these class B liquids which may also prevent vapour escaping.

Additional benefits of using firefighting foams on these liquids are that they cool the liquid surface, reduce the vapour generation rate, obstruct radiation from the flame to the liquid surface and reduce the oxygen level, by the production of steam, where the foam, flame and liquid surface meet.

Lead, as lead tetra-ethyl (or lead tetra-methyl) has been used for more than 60 years to improve the performance (octane rating) of the hydrocarbon mixtures that constitute petrol. However, since 1974, health and environmental concerns have resulted in the progressive reduction in the amounts of lead in petrol. This reduction of the lead content has led to the use of oxygenates, for example ethers and alcohols, as alternative octane improvers. Oxygenates are only used in either leaded or lead-free fuels when the octane rating cannot be achieved cost effectively by refinery processes.

Large scale fire tests have been carried out in the UK to establish whether lead-free petrol, conforming with current British and European standards, would present any problems to the fire service using their standard low expansion foam equipment and techniques (Reference 7). The results showed that providing brigades follow the Minimum Recommended Application Rates given in this Manual, no problems would be expected when using good quality AFFF or FFFP against petrol formulations permitted by current and likely future standards. However, FP gave poor extinction performances against lead-free petrols containing oxygenates although its burnback performances were better than either AFFF or FFFP.

(d)   Water-miscible Class B Liquids

Polar solvents and hydrocarbon liquids that are soluble in water (water-miscible) can dissolve normal firefighting foams. Such liquids include some petrol/alcohol mixtures (gasohol), methyl and ethyl alcohol, acrylonitrile, ethyl acetate, methyl ethyl ketone, acetone, butyl alcohol, isopropyl ether, isopropyl alcohol and many others.

Water-miscible class B liquids, such as some polar solvents, require the use of alcohol resistant type foam concentrates for firefighting and for vapour suppression. These foams form a polymer membrane between the water-miscible and the foam blanket which virtually stops the destruction of the foam and allows vapour suppression and cooling to continue. Alcohol resistant foam concentrates lose effectiveness unless they are applied gently to the surface of polar liquids, avoiding plunging.

6.1.3   Class C Fires

Class C fires are those involving gases or liquefied gases.

In recent years liquefied flammable gases have become an increasingly important source of fuel in commerce and industry. Increased use brings increased transportation of these liquids throughout the country by road, rail, and in UK coastal waters, which in turn increases the possibility of accidental spillage. The product group includes LPG (Liquefied Petroleum Gas, usually propane or butane) liquid ethylene and LNG (Liquefied Natural Gas, i.e. methane).

Boiling points for these liquefied gases are low and so in the event of spillage, rapid vapour production occurs. Due to the greater amounts of vapour produced and the low buoyancy of cold vapour, the dispersal of this vapour is more problematical than from spilled flammable liquids such as petrol. In still air conditions, and where the ground is sloped or channelled, this vapour can travel long distances from its source. Liquefied gas vapour has been known to travel 1,500 metres from a spilled pool of liquid whilst retaining a concentration above the lower flammability limit.

Medium and high expansion foams are suitable for liquefied gas spills both for fire extinguishment and vapour suppression. The surface of the foam in contact with the liquid forms an icy slush which insulates and protects the upper layers of foam, and which in turn acts by reducing the evaporation rate from the liquid. A further important advantage is the relatively low amount of heat transmitted to the liquid by water draining from medium and high expansion foams.

Low expansion foam is not suitable since it increases the rate of evaporation from the liquid. For a liquefied gas spillage any reduction in the rate of evaporation of the liquid is beneficial in that it limits the size of the flammable (or explosive) cloud generated and hence reduces the possibility of ignition.

6.1.4   Class D Fires

Class D fires are those which involve combustible metals such as magnesium, titanium, zirconium, sodium, potassium and lithium. Firefighting foams should not be used with water reactive metals such as sodium and potassium, nor with other water reactive chemicals such as triethyl aluminium and phosphorous pentoxide. Other metal fires are treated as class A fires, but in general the use of media other than foam or water is found to be more suitable.

6.2   Electrical Fires

Firefighting foams are unsuitable for use on fires involving energised electrical equipment. Other extinguishing media are available. Fires in de-energised electrical equipment are treated as either class A or B as appropriate (see above).

6.3   Types of Liquid Fuel Fire

6.3.1   General

The classes of fire discussed in the previous Section have a strong bearing on the tactics and techniques of using firefighting foam. However, the size, shape and general appearance of a fire is also of particular importance when tackling class B or class C fires. Firefighters often refer to spill fires, pool fires and running fires and the variations in firefighting technique required to tackle each. This Section describes these types of fire and how their characteristics can affect the approach to firefighting.

These descriptions relate to ideal conditions which in practice are unlikely to occur exactly as described and in some situations, such as incidents involving aircraft, more than one of these situations may occur simultaneously. Even so, they illustrate the principles involved.

6.3.2 Spill Fires

Spill fires occur in unconfined areas of flammable, or combustible liquids with an average depth of around 25mm or less. There is often variation in the depth of the spill due to unevenness of the surface on which the liquid stands. Because it is unconfined, a spill fire may cover a very large area.

The main characteristic of spill fires is their relatively short burning times. If an average burn rate of 4mm of the depth of fuel per minute is assumed, then most of the fuel involved in a spill fire will have burnt away within 7 minutes of ignition. Such brief burn times are, however, unlikely to occur in practice. Flammable liquid may remain in a ruptured fuel container and burn for a considerable time, continuous leakage may replenish the spill or numerous deep localised burning pools of fuel may form over a large area.

6.3.3 Pool Fires

Pool fires occur in confined pools of flammable, or combustible, liquids which are deeper than 25mm but not as deep as the contents of storage tanks. A pool fire may cover a large area depending on the volume of the fuel source and the area of the confined space. It may take the form of a bunded area in a tank farm or a hollow pit or trench within which flammable liquid has collected from a ruptured process vessel, road or rail tanker.

The difference between pool fires and spill fires is that pools may, depending on depth, continue to burn for a considerable period of time. As a result, firefighters are more likely to encounter a well developed fire burning evenly over a large area, rather than the more isolated, scattered fires which are characteristic of an unconfined spill. Foam may also be subject to more fuel contamination if forceful application is used due to the depth of the fuel. Consequently techniques, such as playing the foam stream against a solid surface and allowing the foam to run onto the fire, may be both desirable and a practical possibility if suitable surfaces are available.

The sustained high levels of heat output may demand more effort to be made in cooling exposed structures both to minimise damage during the fire and to prevent reignition after extinguishment. It should be remembered that if water is used for cooling, it will break down any existing foam blanket in that area, allowing any remaining flames to burn back and preventing further blanket formation until the water application ceases.

The pool fire, therefore, requires a foam with a high fuel tolerance and heat resistance as well as fast flowing characteristics. Adequate post fire security is also required.

6.3.4 Spreading Fires

Spreading fires can be described as unconfined spill or pool fires in which the liquid fuel is being continuously supplemented by a spray, jet or stream from a ruptured tank or equipment. The continuous supply of fuel often results in burning liquid flowing into inaccessible areas, such as drainage systems and floor voids.

An early step in fighting a spreading fire is to stop the flow of product to the flames whenever possible. Water spray provides an excellent screen behind which to approach the fire and close leaking valves for instance. The flow from a storage vessel can also be stopped by water displacement if there is sufficient freeboard above the source of the leak. This method has been successful in the case of a ruptured storage tank line. Water is pumped into the tank to raise the liquid fuel above the level of the outlet line so that water, instead of product, flows from the broken line.

If the flammable liquid is a high flash point fuel, the burn back rate of flames through the spray, jet or stream of fuel leaking from the container may be less than the rate at which the fuel is discharged from the leak. In this situation, the discharging fuel will not be on fire. Consequently, the fire can be extinguished with a foam blanket or water spray in a similar fashion to a pool fire, the only additional precaution being to ensure that the level of fuel does not rise sufficiently to over spill the containment. Sand bagging, diversion channels and pumping out are all useful techniques to help prevent breakdown of containment.

If, on the other hand, the burn back rate of flames through the spray, jet or stream of fuel leaking from the container exceeds the rate at which the fuel is coming out of the container, then the discharging fuel will also be on fire. It may be necessary to use dry powder to extinguish fires in flowing jets of liquid or gas in conjunction with foam application to the spreading fuel. Water sprays are effective in reducing the heat output from burning jets although they will break down any foam blanket already formed.

6.3.5 Running Fires

This term refers to the case when a burning liquid is moving down a slope on a broad front. The situation is rare but extremely hazardous because of the rapidity with which objects and people in the path of the flow can be enveloped. It is not possible to advise any course of action other than rapid evacuation from the oncoming flow. If monitors and hoses are immediately available they could provide sufficiently rapid knockdown.

On some fuels, film-forming foams are considered particularly effective at fast knockdown, although other foams can have similarly rapid effects. Another technique is to lay a band of foam at the lower end of the path of flow so that any pool that builds up will do so beneath a foam blanket. For this type of application fluoroprotein or film-forming alcohol resistant foams might be considered most suitable because of their stability, although other foams would also satisfactorily perform the task.

The main method of combating running fires is by prevention. Firefighters must be aware of any potential for a pool fire to breach or over spill its containment. Firefighting efforts should be adjusted to reduce such a risk, for example, minimising the use of cooling water which could drain into the contained pool and cause overflowing, monitoring the integrity of containing bund walls and evacuating in advance any area which could possibly become inundated.

6.3.6 Other Terms

Various other terms are used for different types of fire and explosion incident such as BLEVE (see Glossary of Terms – Firefighting Foams, at the rear of this Volume), vapour cloud explosion, gas flare, etc. These have not been covered separately since the use of firefighting foam is not directly involved.

Chapter 7 – Application Rates

7.1 General

The application rate of a foam onto a fire is normally expressed as the amount of foam solution, in litres per minute, to be applied to every square metre of the total area to be covered with foam. The following five terms are often used to describe various foam application rates and it is important to know the difference between them, they are:

- Critical Application Rate
- Recommended Minimum Application Rate
- Optimum Application Rate
- Overkill Rate
- Continued Application Rate

The following Sections describe the meanings of these various terms. The most important of these for operational use is **Recommended Minimum Application Rate**.

7.2 Critical Application Rate

The critical application rate is the application rate below which a fire cannot be extinguished. When applied at below this critical rate, the finished foam will be broken down, by both the fuel and the heat of the fire, to such an extent that a complete foam blanket will not be able to form over the surface of the fuel.

7.3 Recommended Minimum Application Rate

7.3.1 General

The Recommended Minimum Application Rate is the minimum rate at which foam solution is recommended to be applied to a fire. The rate assumes that all of the foam made from the foam solution actually reaches the surface of the burning fuel.

The recommended minimum application rate is based on the critical application rate (see above) with an additional 'safety margin' to help to take into account factors such as:

- variations in the quality of foam concentrate;
- variations in the quality of finished foam produced;
- some of the detrimental effects of forceful application.

The Home Office recommended minimum application rates for use by the UK fire service for fires involving water-immiscible class B liquids are given in Section 7.3.2 below. Advice is given concerning the application rates for fires involving water-miscible class B liquids in Section 7.3.4 below.

7.3.2 Fires Involving Water-immiscible Class B Liquids

Tables 7.1 and 7.2 give the minimum application rates of foam solution recommended by the Home Office for use by the UK fire service when using manual firefighting equipment to apply low and medium expansion foam to fires involving water-immiscible class B liquids. Also, recommended durations of foam application are included in the tables.

It should be noted that the figures given in Tables 7.1 and 7.2 relate to **minimum** foam solution application rates and times and assumes that all of the finished foam produced from the foam solution actually reaches the surface of the liquid on fire. These rates should not be considered as being definitive; allowances must be made to compensate for losses due to circumstances such as fall out of finished foam from the foam stream, adverse

Table 7.1: *Home Office Recommended Minimum Application Rates of Foam Solution For the Production of Low Expansion Foam For Use on Liquid Hydrocarbon Fuel (Class B) Fires*

Foam Type	Minimum Application Rate of Foam Solution (lpm/m²)				Minimum Application Time (Minutes)		
	Spill/Bund	Tanks D<45m	Tanks D>=45m D<81m	Tanks D>=81m	Spill	Tanks Fuel Flashpoint >40°C	Tanks Fuel Flashpoint <=40°C /Bund
Protein	6.5	NR	NR	NR	15	NR	NR
Fluoroprotein	5	8.0	9.0	10.0	15	45	60
AFFF	4	6.5	7.3	8.1	15	45	60
FFFP	4	6.5	7.3	8.1	15	45	60
AFFF-AR	4	6.5	7.3	8.1	15	45	60
FFFP-AR	4	6.5	7.3	8.1	15	45	60

Notes to Table 7.1

- < less than
- <= less than or equal to
- \> More than
- \>= more than or equal to
- D Diameter of tank
- lpm/m² litres per minute of foam solution per square metre of burning area of fire
- m metre
- NR Not Recommended for this use

Table 7.2: *Home Office Recommended Minimum Application Rates of Foam Solution For the Production of Medium Expansion Foam For Use on Liquid Hydrocarbon Fuel (Class B) Fires*

Foam Type	Minimum Application Rate of Foam Solution (lpm/m²)	Minimum Application Time (Minutes)	
	Spill/Bund	Spill	Bund
SYNDET	6.5	15	60
Fluoroprotein	5.0	15	60
AFFF	4.0	15	60
FFFP	4.0	15	60
AFFF-AR	4.0	15	60
FFFP-AR	4.0	15	60

Notes to Table 7.2

lpm/m²   litres per minute of foam solution per square metre of burning area of fire weather conditions, breakdown of foam due to flames before it reaches the fuel surface, and loss of foam due to the thermal convection currents caused by the fire. For storage tank fires, these rates need to be increased by up to 60% to account for foam losses.

In addition, it is recommended that application rates should be reviewed if, after 20-30 minutes application, there has been no noticeable reduction in the intensity of the fire.

In practice, the recommended minimum application rates are of great importance in pre-planning the resources needed for a foam attack. It has a direct bearing on the quantity of concentrate, and water required, and also should dictate the amount of delivery equipment, i.e. appliances, monitors, branch pipes, proportioners and hoses.

7.3.3 Fires Involving Water-miscible Class B Liquids

Application rates for water-miscible fuels vary considerably depending on the following factors:

- the type of fuel;
- the depth of fuel;
- the type of foam;
- the manufacturer of the foam;
- the method of foam application.

Some of the most widely used water-miscible liquids include:

> Alcohols (e.g. Methanol, Ethanol, Isopropanol)
> Ketones (e.g. Acetone, Methyl Ethyl Ketone)
> Vinyl Acetate
> Acrylonitrile

Due to the large number of water-miscible fuels in use, and the varying firefighting performance of different foams on each of them, information on the recommended application rates for a particular water-miscible risk should be obtained from the manufacturer of the alcohol resistant foam concentrate to be used.

Typical recommended foam application rates for water-miscible liquid fires range between 4 and 13 litres per minute per square metre. However, it is recommended that the minimum application time for a spill of water-miscible fuel should be 15 minutes and for tanks involving these fuels it should be a minimum of 60 minutes.

On water-miscible liquids, application must be such that the foam blanket is delivered gently onto the liquid surface without submerging the foam or agitating the liquid surface. If some submergence and agitation is unavoidable, the foam blanket will be destroyed at a high rate and much higher application rates and application times will be required.

7.4 Optimum Application Rate

The optimum application rate is sometimes referred to as the most economical rate. It is the rate at which the minimum overall quantity of foam solution is needed to extinguish a fire. This rate usually lies somewhere between the critical application rate and the recommended minimum application rate.

The optimum application rate is not the rate at which the quickest extinction is achieved. To achieve the quickest extinction time, rates in excess of the optimum application rate are required. However, the small reductions in extinction times achieved by these increased application rates are at the cost of large increases in the use of resources such as water, foam concentrate etc. For some applications, such as those involving air crashes, quick extinction times are of the utmost priority and can be considered a worthwhile use of these resources.

7.5 Overkill Rate

There is a limit to how quickly a fire can be extinguished when using firefighting foam. Once the application rate has reached a certain level, higher application rates give no improvements in extinction time, they only result in a wastage of resources. These higher application rates are known as overkill rates.

7.6 Continued Application Rate

Various standards quote lower rates for continued application after a fire situation has been extinguished. These rates should be sufficient to maintain the integrity of the foam blanket and are often around 50% of the minimum recommended foam application rate.

Firefighting Foam – Technical

References

1. CFBAC, JCFR Report 19, *Trials of Medium and High Expansion Foams on Petrol Fires*, P L Parsons, 1982.

2. SRDB Publication 12/90, *Chemical Effects of Additives on Fire Appliances and Associated Equipment*, B P Johnson, 1990.

3. CFBAC, JCFR Report 31, *Additives for Hosereel Systems Trials of Foams on 40m^2 Petrol Fires*, J A Foster, 1988.

4. CFBAC, JCFR Report 79, *Class A Additives*, K Bosley, 1997.

5. FRDG Publication 2/93, *A Comparison of Various Foams when used against Large Scale Petroleum Fires*, B P Johnson, 1993. ISBN 0-86252-949-2

6. Foundation For Water Research, R&D Note 339, *A Review of Fire Fighting Foams to Identify Priorities For EQS Development*.

7. CFBAC, JCFR Report 49, *The Use of Foam Against Large-Scale Petroleum Fires Involving Lead-Free Petrol Summary Report*, J A Foster, 1992.

8. CFBAC, JCFR Report 43, *Equipment For The Induction Of Additives Into Hose Reel Systems*, J A Foster, B P Johnson, 1991. ISBN 0-86252-652-3

9. FRDG Publication 1/94, *A Brief Assessment of a Compressed Air Foam System*, M D Thomas, 1994. ISBN 1-85893-149-5

10. FRDG Publication 3/91, *Additives for Hosereel Systems: Trials of Foam on Wooden Crib Fires*, B P Johnson, 1991.

Firefighting Foam – Technical

Further Reading

1. CFBAC, JCFR Report 40, *Survey of Firefighting Foams, Associated Equipment and Tactics* [Ewbank Preece Reports] 1990. ISBN 0 82652 556 X
   Part 1 : Firefighting Foams
   Part 2 : Tactics and Equipment
   Part 3 : Large Tank Fires

2. Fire Service Manual – Volume 2 – Fire Service Operations – Petrochemicals.

3. Fire Service Manual – Volume 2 – Fire Service Operations – Firefighting Foam.

4. CFBAC, JCFR Report 46, *Additives for Hosereel Systems Summary Report*, B P Johnson, 1992.

5. CFBAC, JCFR Report 48, *An Assessment of the Damage to Tank Farms in Kuwait Following Hostilities and their Implications for UK Practice Summary Report*, M W Freeman, 1992.

6. SRDB Publication 9/87, *Pilot Study on Low Expansion Foam Making Branchpipes*, B P Johnson, P L Parsons, 1987.

7. SRDB Publication 22/88, *Additives for Hosereel Systems: Preliminary Trials of Foam on Small Scale Isopropanol Fires*, B P Johnson, 1988.

8. FRDG Publication 5/91, *Additives for Hosereel Systems: Trials of Foam On Tyre Fires*, B P Johnson, 1991.

9. FRDG Publication 4/94, *A Comparison Of Various Low Expansion Foams When Used Against The Proposed ISO And CEN Standard Medium Scale ($4.5M^2$) Hydrocarbon Fuel Test Fire*, B P Johnson, 1994

Firefighting Foam – Technical

Glossary of Terms: Firefighting Foams

(Note: Not all of these terms have been used in this Manual of Firemanship but they have been included here for completeness)

Accelerated ageing	Storage of foam concentrate at high temperatures to indicate long term storage properties of the foam concentrate at ambient temperatures.
Acidity	See pH.
Alcohol resistant foam concentrates	These may be suitable for use on hydrocarbon fuels, and additionally are resistant to breakdown when applied to the surface of water-miscible liquid fuels. Some alcohol resistant foam concentrates may precipitate a polymeric membrane on the surface of water-miscible liquid fuels.
Alkalinity	See pH.
Application rate	The rate at which a foam solution is applied to a fire. Usually expressed as litres of foam solution per square metre of the fire surface area per minute (lpm/m^2).
AFFF concentrate	Aqueous film-forming foam. AFFFs are generally based on mixtures of hydrocarbon and fluorinated surface active agents and have the ability to form an aqueous film on the surface of *some* hydrocarbon fuels.
Aspiration	The addition or entrainment of air into foam solution.
Aspirated foam	Foam that is made when foam solution is passed through purpose designed foam-making equipment, such as a foam-making branch. These mix in air (aspirate) and then agitate the mixture sufficiently to produce finished foam. (see also primary aspirated foam and secondary aspirated foam).
Base injection (Subsurface injection)	The introduction of fuel-tolerant primary aspirated finished foam beneath the surface of certain flammable and combustible hydrocarbons, to effect fire extinguishment. Usually used for the protection of fixed roof hydrocarbon fuel storage tanks.
Bite	The formation of an initial area of foam blanket on the surface of a burning liquid fuel.

Boiling liquid expanding vapour explosion (BLEVE)	The catastrophic failure of a tank containing pressure liquefied gas (PLG) due to mechanical damage or adverse heat exposure will result in a BLEVE. A BLEVE will produce blast and projectile hazards. If the contents of the tank are toxic, then health and exposure hazards may occur. If the contents are flammable, then a fireball may occur with associated thermal radiation and fire engulfment hazards.
Boil-over	Violent ejection of flammable liquid from its container, caused by vaporisation of a water layer beneath the body of the liquid. It will generally only occur after a lengthy burning period in wide flashpoint range products, such as crude oil. The water layer may already have been in the container before the fire began or may be the result of the inadvertent application of water (perhaps during cooling of the container walls), or from the drainage of foam solution from finished foam applied to the fire. (see also froth-over and slop-over).
Bund area (Dike area)	An area surrounding a storage tank which is designed to contain the liquid product in the event of a tank rupture.
Branch	A hand-held foam maker and nozzle.
Burnback resistance	The ability of a foam blanket to resist direct flame and heat impingement.
Candling	Refers to the thin intermittent flames that can move over the surface of a foam blanket even after the main liquid fuel fire has been extinguished.
Chemical foam	A finished foam produced by mixing two or more chemicals. The bubbles are typically caused by carbon dioxide released by the reaction.
Classes of Fire	In the UK the standard classification of fire types is defined in BS EN 2: 1992 as follows: 'Class A: fires involving solid materials, usually of an organic nature, in which combustion normally takes place with the formation of glowing embers. Class B: fires involving liquids or liquefiable solids. Class C: fires involving gases. Class D: fires involving metals.' Electrical fires are not included in this system of classification.
Cloud point	The lowest temperature at which a liquid remains clear. Usually only applicable to high expansion foam concentrates.
Combustible liquid	Any liquid having a flashpoint at or above 37.8°C (100°F).

Concentration

To achieve effective performance, foam concentrates must be mixed to the concentration recommended by the manufacturer. For each 100 litres of the required foam solution, the foam concentrate must be mixed as follows:

Recommended Concentration	Volume of Foam Concentrate (litres)	Volume of Water (litres)	Volume of Foam Solution (litres)
1%	1	99	100
3%	3	97	100
6%	6	94	100

Critical application rate

The foam application rate below which a fire cannot be extinguished.

Crude oil

Petroleum, in its natural state, as extracted from the earth. Consequently, there are many different types of crude oil, each with different characteristics and each yielding different quality products. The various constituents ensure that crude oils generally have wide ranging flash points with usually sufficient fractions (or light ends) to classify them as class A petroleum products.

Density

The mass per unit volume of a material:

$$\text{Density} = \frac{\text{mass}}{\text{volume}}$$

Dike area

See Bund area.

Discharge rate (high expansion foam)

The discharge rate of a high expansion foam generator measured in cubic metres/min (m^3/min) of foam at a stated expansion ratio.

Drainage time

The time taken for a percentage of the liquid content of a finished foam sample of a stated depth to drain out of the foam. For low expansion foam, times taken for 25% of the foam solution to drain out are usually given; for medium and high expansion foams 50% drainage times are usually given.

Expansion ratio

The ratio of the total volume of finished foam to the volume of foam solution used to produce it:

$$\text{Expansion ratio} = \frac{\text{volume of finished foam}}{\text{volume of foam solution used to produce it}}$$

Film-forming

A finished foam, foam solution or foam concentrate that forms a spreading, thin, aqueous film on the surface of some hydrocarbon liquids.

FFFP foam concentrates

Film-forming fluoroprotein. These are fluoroprotein foam concentrates which have the ability to form an aqueous film on the surface of some hydrocarbon fuels.

Finished foam	The foam as applied to the fire. It will consist of a mixture of foam solution that has been mixed with air. The foam may be primary aspirated or secondary aspirated.
Flammable liquid	Any liquid having a flashpoint below 37.8°C (100°F).
Flashback	The re-ignition of a flammable liquid caused by the exposure of its vapour to a source of ignition such as a hot metal surface or a spark.
Flashpoint	The lowest temperature at which a flame can propagate in the vapour above a liquid.
Flow requirement (low and medium expansion)	The nominal supply rate of foam solution required by a foam branch, measured in litres per minute.
Fluoroprotein (FP) foam concentrate	A hydrolysed protein based foam concentrate with added fluorinated surface active agents.
Foam	The result of mixing foam concentrates, water and air to produce bubbles.
Foam concentrate	The foam as supplied by the manufacturer in liquid form; this is sometimes referred to as 'foam compound', 'foam liquid' or by trade or brand names.
Foam, dry	Foam with a long drainage time, i.e. the liquid content of the foam takes a long period of time to drain out of the foam; the foam is very stable.
Foam generator (high expansion)	A mechanical device in which foam solution is sprayed onto a net screen through which air is being forced by a fan.
Foam generator (low expansion)	Similar to a foam-making branch, but inserted in a line of hose so that the finished foam passes along the hose to a discharge nozzle.
Foam-making branch (foam-making branchpipe, FMB)	The equipment by which the foam solution is normally mixed with air and delivered to the fire as a finished foam.
Foam monitor	A larger version of a foam-making branch which cannot be hand-held.
Foam solution	A well mixed solution of foam concentrate in water at the appropriate concentration.
Foam, wet	Foam with a short drainage time, i.e. the liquid content of the foam takes a short period of time to drain out of the foam; the foam breaks down quickly.

Freeze point	The highest temperature at which a material can exist as a solid.
Froth-over	Overflow of a non-burning flammable liquid from a container due to the thermal expansion of the liquid or violent boiling on top of and within the upper layers of the liquid due to the presence of small quantities of water. (see also boil-over and slop-over)
Hazmat	A proprietary trade name used to describe special types of foam which can be used to suppress the vapour production of certain hazardous materials (toxic, odorous and/or flammable).
Heat resistance	The ability of a foam blanket to withstand the effects of exposure to heat.
High expansion foam (HX)	Finished foam of expansion ratio greater than 200:1
Hydrocarbon fuel	Fuels based exclusively on chains or rings of linked hydrogen and carbon atoms. Hydrocarbon fuels are not miscible with water.
Induction	The entrainment of foam concentrate into the water stream.
Inductor (Eductor)	A device used to introduce foam concentrate into a water line.
Induction rate (pick-up rate)	The percentage at which foam concentrate is proportioned in to water by an inductor in order to produce a foam solution. Normally this is 1%, 3% or 6%.
Inline inductor	An inductor inserted in to a hose line in order to induce foam concentrate prior to the water reaching the foam-making branch.
Knockdown	The ability of a foam to quickly control flames. Knockdown does not necessarily mean extinguishment.
Low expansion foam (LX)	Finished foam of expansion ratio of less than or equal to 20:1.
Mechanical foam	Foam produced by a physical agitation of a mixture of water, foam concentrate and air.
Medium expansion foam (MX)	Finished foam of expansion ratio greater than 20:1. but less than or equal to 200:1.
Minimum use temperature	The lowest temperature at which the foam concentrate can be used at the correct concentration through conventional equipment such as inline inductors and other proportioning devices.
Monitor	A large throughput branch (water or foam-making) which is normally mounted on a vehicle, trailer or on a fixed or portable pedestal.

Multipurpose foam concentrates	Another name given to alcohol resistant foam concentrates.
Newtonian liquids	The viscosity of Newtonian liquids remains the same no matter how quickly or slowly they are flowing (see also non-Newtonian pseudo-plastic liquids). Most non-alcohol resistant foam concentrates (such as AFFF, FFFP, FP, P and SYNDET) are Newtonian liquids.
Non-aspirated (Unaspirated)	The application, by any appropriate means, of a firefighting liquid that does not mix the liquid with air to produce foam (i.e. aspiration does not occur). The term 'non-aspirated foam' is often used incorrectly to describe the product of a foam solution that has been passed through equipment that has not been specifically designed to produce foam, such as a water branch. However, the use of this type of equipment will often result in some aspiration of a foam solution. This is because air is usually entrained into a jet or spray of foam solution as it leaves the branch, as it travels through the air due to the turbulence produced by the stream and/or when it strikes an object. This causes further turbulence and air mixing. There is sufficient air entrained by these processes to produce a foam of very low expansion (often with an expansion ratio of less than 5:1). Consequently, the term secondary aspirated foam is preferred in these cases (see also primary aspirated and secondary aspirated foam).
Non-Newtonian pseudo-plastic liquids	As the rate of flow of non-Newtonian pseudo-plastic liquids increases, their viscosity decreases and so they flow more easily. Consequently, getting them to flow initially can be difficult, but once flowing, their viscosity reduces to a more acceptable level. Many alcohol resistant foam concentrates (such as AFFF-AR and FFFP-AR) are considered to be non-Newtonian pseudo-plastic liquids.
Oleophobic	Oil repellent.
Over-the-top foam application	The application of foam by projecting it over the sides of a storage tank and directly on to the surface of the contained fuel.
pH (Acidity/Alkalinity)	Measurement of the acidity to alkalinity of a liquid on a scale of 1 to 14. A pH of 7 is neutral (like that of pure water), a pH of 1 is very acidic, a pH of 14 is very alkaline.
Polar solvent	This term is generally used to describe any liquid which destroys standard foams, although it actually refers to liquids whose molecules possess a permanent dielectric discharge e.g. Alcohols, ketones. Most polar solvents are water-miscible.
Pour point	The lowest temperature at which a foam concentrate is fluid enough to pour. This is generally a few degrees above its freezing point.

Preburn time The time between ignition of a fire and the commencement of foam application.

Premix solution A mixture in correct proportions of a foam concentrate and water. Use of this term generally implies that the foam is stored in a premix form, as in a portable foam fire extinguisher or as foam solution in a fire appliance water tank.

Primary aspirated foam Finished foam produced from foam solutions that are passed through purpose designed foam-making equipment. (See secondary aspirated foam).

Proportioner A device where foam concentrate and water are mixed to form a foam solution.

Protein (P) foam concentrate Protein foam concentrate contains organic concentrates derived from natural vegetable or animal sources. Hydrolysed products of protein provide exceptionally stable and heat resistant properties to foams although they lack fuel tolerance and have slow knock-down performance.

Relative density see Specific gravity

Secondary aspirated foam Finished foams that are produced from foam solutions that are applied other than by purpose designed foam-making equipment, usually standard water devices. (See primary aspirated foam).

Security The ability of a foam to seal around hot objects and prevent reignition.

Shear strength The measurement of the stiffness of a finished foam sample when measured with a foam viscometer. Units of measurement are Newtons per square metre (n/m^2).

Slop-over When some burning liquids, such as heavy fuel oils or crude oils, become extremely hot, any applied water may begin to boil on contact with the fuel, the resulting rapid expansion as it converts to steam may cause burning fuel to overflow its containment and the fire to spread (see also boil-over and froth-over).

Solution transit time The time taken for foam solution to pass from the point where foam concentrate is introduced in to the water stream to when finished foam is produced.

Specific gravity The specific gravity of a material is a measure of the density of the material in relation to the density of water. The specific gravity is calculated as:-

$$\text{Specific Gravity} = \frac{\text{Density of material}}{\text{Density of water}}$$

A liquid with a specific gravity of less than one will float on water (unless it is water-miscible); a specific gravity of more than one indicates that water will float on top of the liquid.

Spill fire A flammable liquid fire having an average depth of not more than 25mm.

Stability The ability of a finished foam to retain shape and form particularly in the presence of heat, flame and/or other liquids. The 25% drainage time is often used as a measure for stability.

Subsurface injection See base injection.

Surface active agents A chemical ingredient of some foam concentrates. Finished foams is stabilised by the addition of surface active agents (or surfactants) which promote air/water stability by reducing the liquids surface tension. Most surface active agents are organic in nature and common examples are soaps and detergents.

Synthetic detergent (SYNDET) foam concentrate These are based upon mixtures of hydrocarbon surface active agents and may contain fluorinated surface active agents with additional stabilisers. They are multipurpose foams in that they can be used at low, medium and high expansion.

Venturi A constricted portion of a pipe or tube which will increase water velocity, thus momentarily reducing its pressure. It is in this reduced pressure that foam concentrate is introduced. The pressure difference across the venturi can be used to force foam concentrate into the water.

Viscosity This is a measure of how well a liquid will flow. Liquids are generally classed as either being non-Newtonian or Newtonian. A low viscosity is often desirable because it improves the flow characteristics of a foam concentrate through pick-up tubes, pipework and induction equipment.

Viscosity will also vary with foam concentrate type and with concentration. AFFF foam concentrates at 3% and 6% oncentrations tend to be the least viscous, closely followed by P, FP and FFFP foam concentrates at 6%. AFFF at 1% and SYNDET foams, P, FP and FFFP foam concentrates at 3% concentration are appreciably more viscous than these. The alcohol resistant foams are often the most viscous although recent developments have dramatically reduced the viscosity of some products.

In addition, the viscosity of all foam concentrates will vary with temperature and may be affected by the age of the foam concentrate. Manufacturers often state the viscosity of their products when measured at 20°C; lower temperatures will result in higher viscosity.

Water-immiscible liquid A liquid that is not soluble in water.

Water-miscible liquid A liquid that is soluble in water. Polar solvents and hydrocarbon liquids that are water-miscible can dissolve normal firefighting foams (see also alcohol resistant foam concentrates).

Wetting agent A chemical compound which, when added to water in correct proportions, materially reduces its surface tension, increases its penetrating and spreading abilities and may also provide foaming characteristics.